AF588518

Is a coincidence of coincidences just a coincidence? Following Edward A Wilson into the unknown.

Horatio. "O day and night, but this is wondrous strange!"

Hamlet. "And therefore as a stranger give it welcome.
There are more things in heaven and earth, Horatio,
Than are dreamt of in your philosophy."

(Shakespeare, Hamlet, Act 1 Scene 5)

Published by
Reardon Publishing
PO Box 919, Cheltenham, GL50 9AN, England
antarcticbookshop.com

Written by
John E.C. Flux

isbn: 9781901037272

Book Design by
Nicholas Reardon

Cover Photos:
Front: Wilson's treecreeper *(Certhia familiaris)* painting discovered in Antarctica in 2016.
Rear: Author's flashlight photo of them sleeping in the soft bark of Redwood trees *(Sequoia gigantea)*

Foreword

This booklet is a totally new field for me. I had no interest in Science-fiction, fortune-tellers, ghosts, aliens from space, or UFOs. I was an ecologist of the old school, a naturalist content to observe, record, describe, understand, and explain real facts: life, on this planet now, and how everything interacts, including humanity. As a schoolboy I read J. W. Dunne's (1927) book “An Experiment with Time”, which deals convincingly with dreams that foretold future events; but nothing I recorded in a notebook for a few months happened, and I abandoned the topic.

Yet there are clearly things about the human mind that are still unknown “magic”and have no scientific explanation. A hypnotist can remove viral warts by telling a patient to make them disappear on one part of the body, and they do. If told a cold pencil applied to a patient's arm is red hot, the patient develops a typical burn blister (Lewis Thomas, “Late Night Thoughts”, 1985). Neither the method of thought transfer, nor the ability of the body to produce the right response, can be understood at present. Other unknowns that Thomas lists are “...what goes on in the mind of a honey bee” and “...the problem of music – what music is, why is it indispensable for human existence...” (Recent studies show that pleasant music reduces pain, but the mechanism is still unclear.) C.H. Waddington “The Nature of Life” (1963) concludes: “Since the nature of self-awareness completely resists our understanding...the very fact of existence as we know it in our experience is essentially a mystery”. The ability of autistic savants to immediately tell ones weekday of birth given date of birth, instantly list prime numbers, and communicate with animals, defy present understanding (Grandin and Johnson 2006). One savant, Daniel Tammet, explained how he multiplied two three-digit numbers: “When I multiply numbers together, I see two shapes. The image starts to change and evolve, and a third shape emerges. That's the answer.” (Richard Johnson, The Guardian, 12 February 2005.)

The difference between people in the clarity and range of their senses is an allied problem. It is not possible to explain to a red/green colour-blind person what “red” looks like, nor to understand what colour the rare person with UV vision receptors sees. This must confuse research on synaesthesia - the mental vision of numbers as coloured, or the blending of other sensory information; words with flavours, or sounds with textures (K. Sukel, 10 June 2023, New Scientist). Many people cannot taste phenylthiourea; mace may or may not taste like nutmeg.

Freesia and boronia flowers are highly scented to most people, but unscented to others. Genetic differences are responsible, but it is useless to say they smell or taste like something else because we cannot tell what anyone else experiences.

Coincidences seemed to be another problem to add to the list, as defined in the Concise OED: “notable concurrence of events, or circumstances without apparent causal connection”.

To try to explore coincidences I read more widely. I ordered a copy of a psychologist's analysis of the forms they take (Beitman 2022), and in the meantime came across Richard Bach's “Illusions; the adventures of a reluctant Messiah” (1977) in my local bookshop. On p.62 it said “...hold some problem in your mind, then open any book handy and see what it tells you...” So I did; shut my eyes, walked to a bookcase, and opened a book.

(By chance, at dawn on 6 February 2024, sunlight on the bookcase let me arrange my shadow.
It is a coincidence that placing my hand central put my fingers on Macan's white book, and the big brown volume, central on the bottom shelf, is Wilson's Grouse Report).

The page was a fascinating account of factors controlling the temperature of lakes and rivers, and I never knew that freshwater entering a salt-water lake formed a surface layer preventing heat loss, so deeper water could reach 50-71°C. My interest in climate change gained a new dimension, especially the danger of sea-level rise following warm sea water undercutting Thwaites Glacier (Douglas Fox, 2022, "The coming collapse", Scientific American). So what was the book? I shut it to read "Freshwater Ecology 2nd Edition T.T. Macan," (1974); and Macan had lectured to us at the Windermere student course on 9 March 1956. His foreword was remarkably humorous, relevant, and exactly what I needed:

> "A foreword is used in different ways by different authors. Some write it with the fire and fervour of one preaching a new religion, implying freely that those who do not follow their ideas will be cast into outer darkness by the scientific fraternity in this world and probably also by whoever arranges such things in the next. The only doctrine I wish to preach is toleration..." Amen to that.

Again by coincidence, I came across John Locke's appeal for "Toleration... the chief... mark of the True Church" on behalf of Thomas Aikenhead, a Scottish divinity student aged 18, who favoured pantheism and had suggested the biblical account of Creation was a myth: but Aikenhead was hanged on 8 January 1697 (Arthur Herman, 2001, "The Scottish Enlightenment: The Scots' invention of the modern world"). The support of such religious doctrines is another mystery. As Philip Ball warns: "History makes it clear that religion need not be anti-intellectual, anti-science, anti-democratic, anti-humanist. But it also shows that it can be." (Philip Ball, 2017, Demographics. In Jim Al-Khalili, Ed., "What's Next?"). It played a major role in colonial subjugation: "They came with a Bible and their religion; stole our land, crushed our spirit, and now tell us we should be thankful to the 'Lord' for being saved." Chief Pontiac, c1750. However, some Christians, often enlightened by studying the faith of indigenous races, promote the view that "everything created is holy and to be revered" (Anne Rowthorn, 1989, "Caring for Creation: Toward an ethic of responsibility").

Bernard Beitman's book “Meaningful Coincidences: How and why synchronicity and serendipity happen” then arrived, and I found it a useful and scientifically accurate analysis of the frequency of types of coincidences, and their classification. To quote him: “Humans seek coherent structure and order. We seek patterns by which to describe, predict, and control realities... Coincidences are formed from two kinds of events: **mind and object.** Mind events ... are primarily private events...like grief, and sensations, like pain...**Object** events occur in the public sphere so that someone else could possibly observe them.” His classification lists **mind-object** (the commonest form), **mind-mind, object-object,** and my problem: “**Meta-coincidences** are coincidences about coincidences”. Another useful overview is “Harper's Encyclopedia of Mystical & Paranormal Experience” by R. E. Guiley, 1991.

Many scientists spurn these mystic investigations, and this essay will doubtless evoke their criticism. I take refuge in my University of Aberdeen Marischal College motto “Thai Haif Said: Quhat Say Thay: Lat Thame Say”; English translation “They have said. What say they? Let them say!” Wilson may well have known the Latin version, *“Aiunt. Quid aiunt? Aiant.”* In brief, ignore unfriendly critics. As Lao Tsu warned, “Care about what other people think and you will always be their prisoner”.

Being a loner is a great help: “Loneliness is the safest place I know” (Edgar Allen Poe). And as Jean-Jacques Rousseau eventually found when spurned for his published theories by both critics and all his friends: “...being certain that there is no new hold which they can use to inflict some permanent suffering on me, I laugh at all their scheming and enjoy my own existence in spite of them.” (Rousseau, 1782, “Meditations of a solitary walker”).

Introduction

It began with a newspaper article, which earned "Highly commended" in the Prinz Awards of 2018:

"In 2016, Antarctic Heritage Trust conservators [actually Conservator Josefin Bergmark-Jimenez in Borchgrevink's hut at Cape Adair] made a surprising discovery – a painting that had lain forgotten in Antarctica for more than a century. This was an opportunity to raise the Trust's profile. But given its small budget for the project, a clash in the release timing with the British election and some doubt as to the significance of the find, the Communications Strategy had to be simple, smart and effective.
It was. The story was run by 218 news outlets in 32 countries."

The newly discovered painting was of a treecreeper (probably *Certhia familiaris* from its Swiss alpine location, long toenails, and white eye-stripe) by Edward Wilson. Kennedy Warne wrote about it for the New Zealand Geographic magazine (2017), ending with:

"Back to the bird: why did Wilson take a painting he had made a decade earlier in Switzerland with him to the ends of the earth? And how did it get to Cape Adair? Wilson was in Scott's main party, based at Cape Evans, 700 kilometers south of the northern party at Cape Adair. Was it a gift to someone in that party? We may never know. Perhaps there was simply something about that painting that spoke to Wilson. A dead songbird on a snowy white background. Beautiful for all its sorrow."

I read about it in our local newspaper, The Dominion Post, on 13 June 2017. It seemed obvious to me that the (unknown) reason Dr Edward Wilson had taken the painting of a dead treecreeper [Fig.1W] to Antarctica on Scott's polar expedition was to remind him of his home in England at an estate called The Crippetts: just as I had carried a photograph of a sleeping treecreeper [1F] in my wallet for 7 years after emigrating to New Zealand from Scotland in 1960. Mine was a memento of Dunnottar woods in Stonehaven (near Aberdeen), where I spent much of my boyhood watching birds on my own.

The coincidence that we had chosen the same – rather unusual – bird intrigued me. I knew of Wilson's illustrations of mountain hare colour change in Barrett-Hamilton (1912) "A History of British Mammals" [2W] because my PhD at Aberdeen University was on the ecology of mountain hares [2F]. But I was unaware that he had drawn Red Grouse flank feathers [3W] and toenail grooves [4W] to sex and age the birds in "The report of the committee of enquiry on grouse disease" (Leslie and Shipley, 1912), when I published the same things [3F,4F] in "The Scottish Landowner" (1958) – at the insistence of the supervisor of my student vacation job (examining 5000 shot grouse). All scientists acknowledge earlier work they are aware of.

Following this, I started collecting books about Wilson, and found that his illustrations kept reminding me of sketches or photographs I had made years ago. Especially valuable sources were D.M. and C.J. Wilson's "Edward Wilson's Nature Notebooks" (2004) and "Edward Wilson's Antarctic Notebooks" (2011). All the following paired images are independent; none were taken to copy a Wilson image, and many are from school or university notebooks drawn before I had even heard of him. Several have two or more points of similarity, often including unusual small details, as noted in their captions. No photos in this booklet have been 'photo-shopped' in any way.

Selection of illustrations for comparison

Apart from the first four introductory pairs, there is no attempt to arrange topics in any order. Every new book of Wilson's art I found produced more to include, so after five years I decided to stop at 150 random pairs.

Number 1.

1. Wilson's treecreeper *(Certhia familiaris)* painting discovered in Antarctica in 2016 (© AHT), and my flashlight photo of one sleeping on 27.12.1954.
They dig hollows in the soft bark of Redwood trees *(Sequoia gigantea)* to sleep in, so only their backs and stiff tails show.

Number 2.

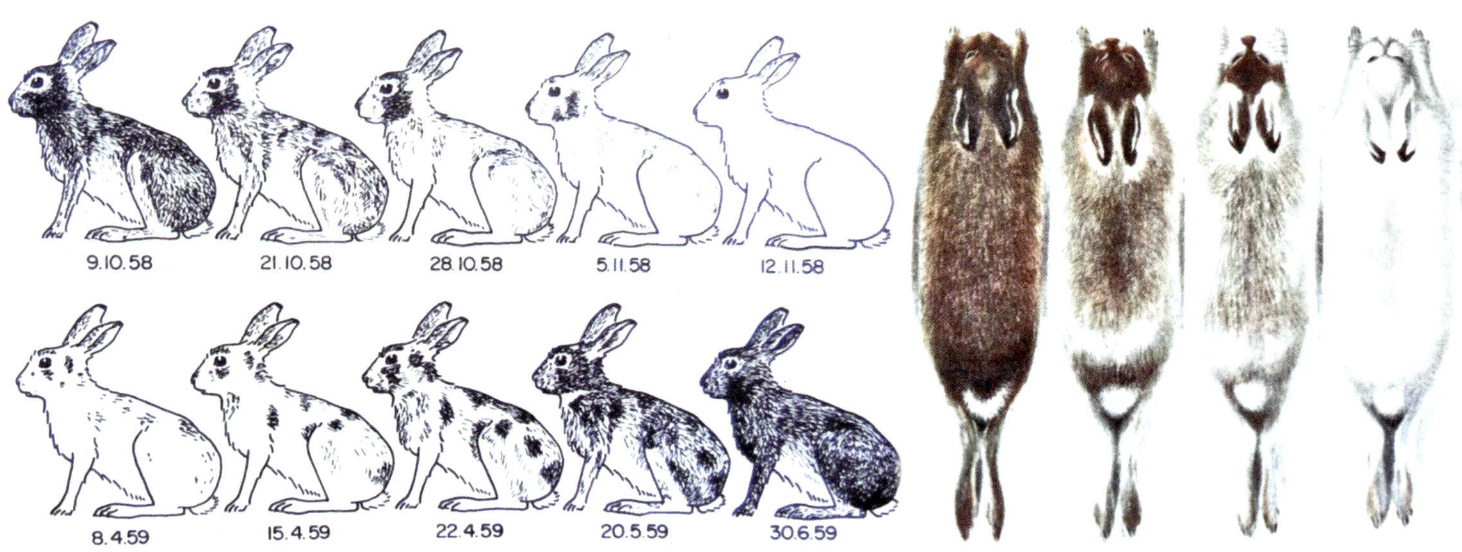

2. I studied colour change in mountain hares on Morven,
NE Scotland, in 1957-59,
and knew Wilson's illustration in "A History of British Mammals"
by Barrett-Hamilton (1912).

Number 3.

Summer flank feathers of a hen grouse (top left) and of a cock (top right). Winter flank feathers of a hen (bottom left) and of a cock (bottom right). Note the more barred effect in the hen.

Female Grouse, Red Type: Feathers from flanks.

3. Red Grouse *(Lagopus scoticus)* flank feathers, used to age and sex the birds.

Number 4.

On the left is the claw of a young grouse. The other two are the claws from old birds: the upper one has the previous year's claw still attached, the lower shows the scar after moulting.

4. Grouse toe nails. My article "Red grouse in autumn: age and sex characteristics studied" was published in Scottish Landowner in 1958. I had not seen Wilson's paintings in "The grouse in health and disease" by Lesley and Shipley (1912).

Number 5.

5. Swift *(Apus apus)* eye wind-shield in my 1949 notebook, and Wilson's version.

Number 6.

6. My sketches of a dead magpie *(Pica pica)*,
and Wilson's with colour notes.

Number 7.

7. Dead sparrowhawks *(Accipiter nisus);* both with one foot closed, one half open. My drawing of a bird's head and feet is typical of many of Wilson's paintings.

Number 8.

8. My only swallow *(Hirundo rustica)* photographed from below, taken in India, was on a red and yellow perch, as Wilson's is. In a search online of "swallows", only 13 of 3,985 were taken from below (all on normal telephone lines).

Number 9.

9. Grey wagtails *(Motacilla cineria)* on red stones. Mine in north Scotland is on Old Red Sandstone, Wilson's possibly on Devonian sandstone.

Number 10.

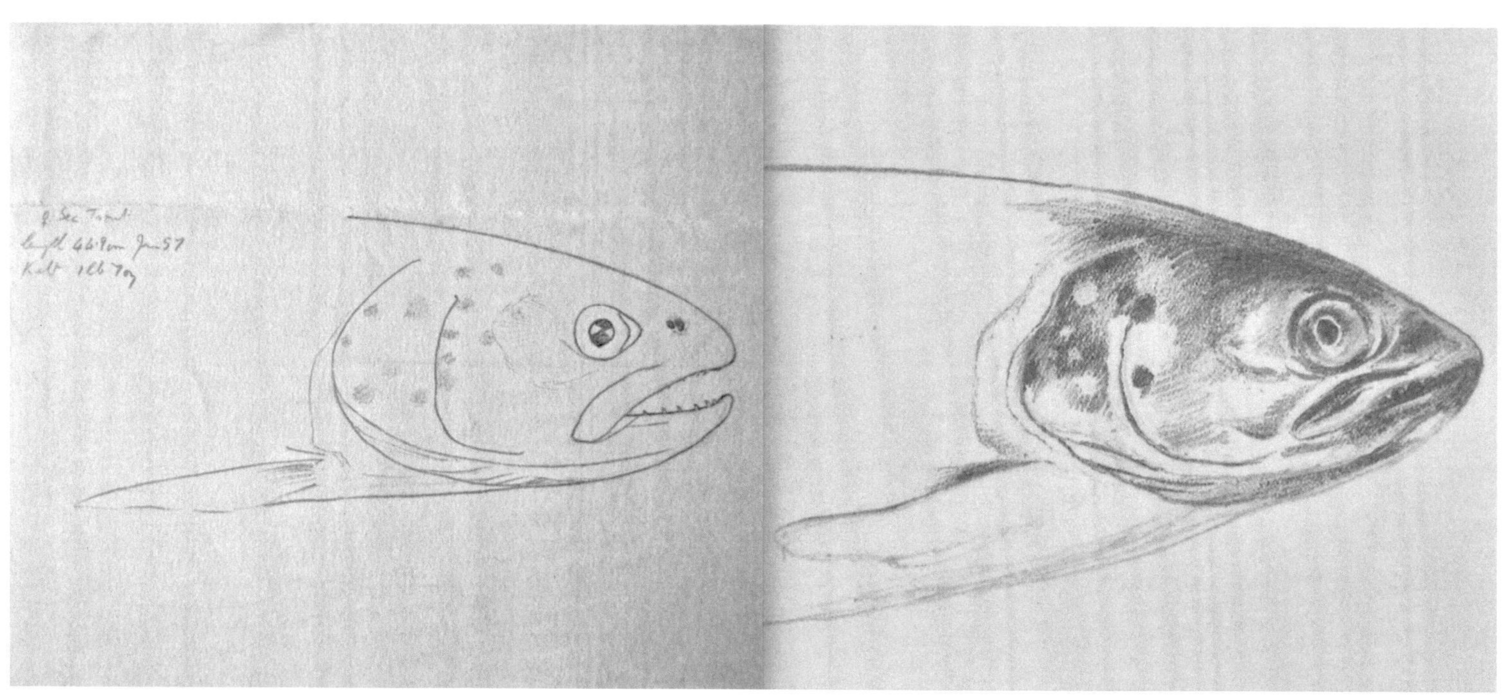

10. Sea trout *(Salmo trutta)* heads; both front part only.

Number 11.

11. We both painted/photographed many freshwater fish, so matching is common: "...char *(Salvelinus alpinus)* the prettiest fish ever invented" Wilson wrote.

Number 12.

12. But demonstrating the way fish jaws work is more unusual.

Number 13.

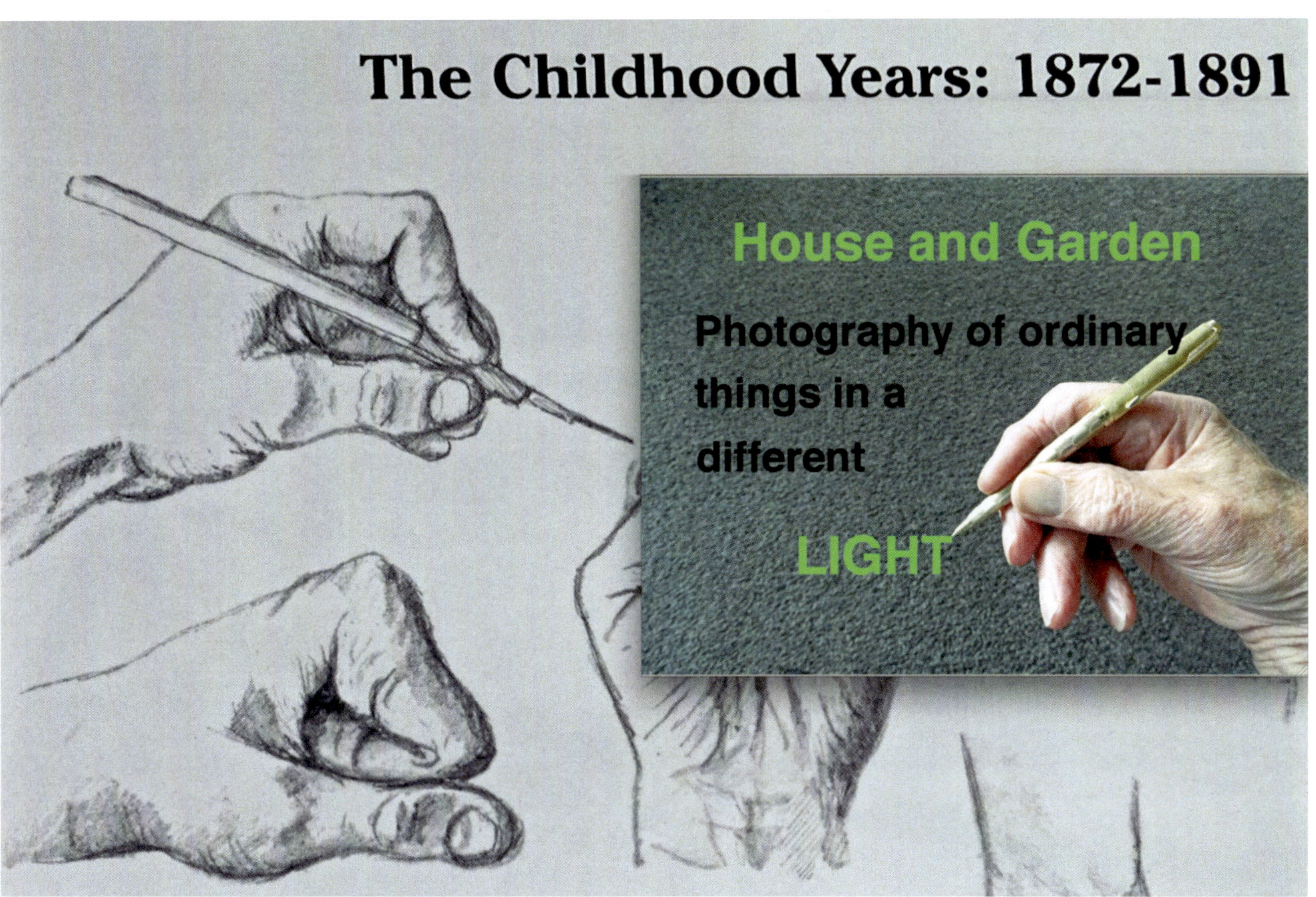

13. I used my pen to introduce a talk on photography before seeing Wilson's version.

Number 14.

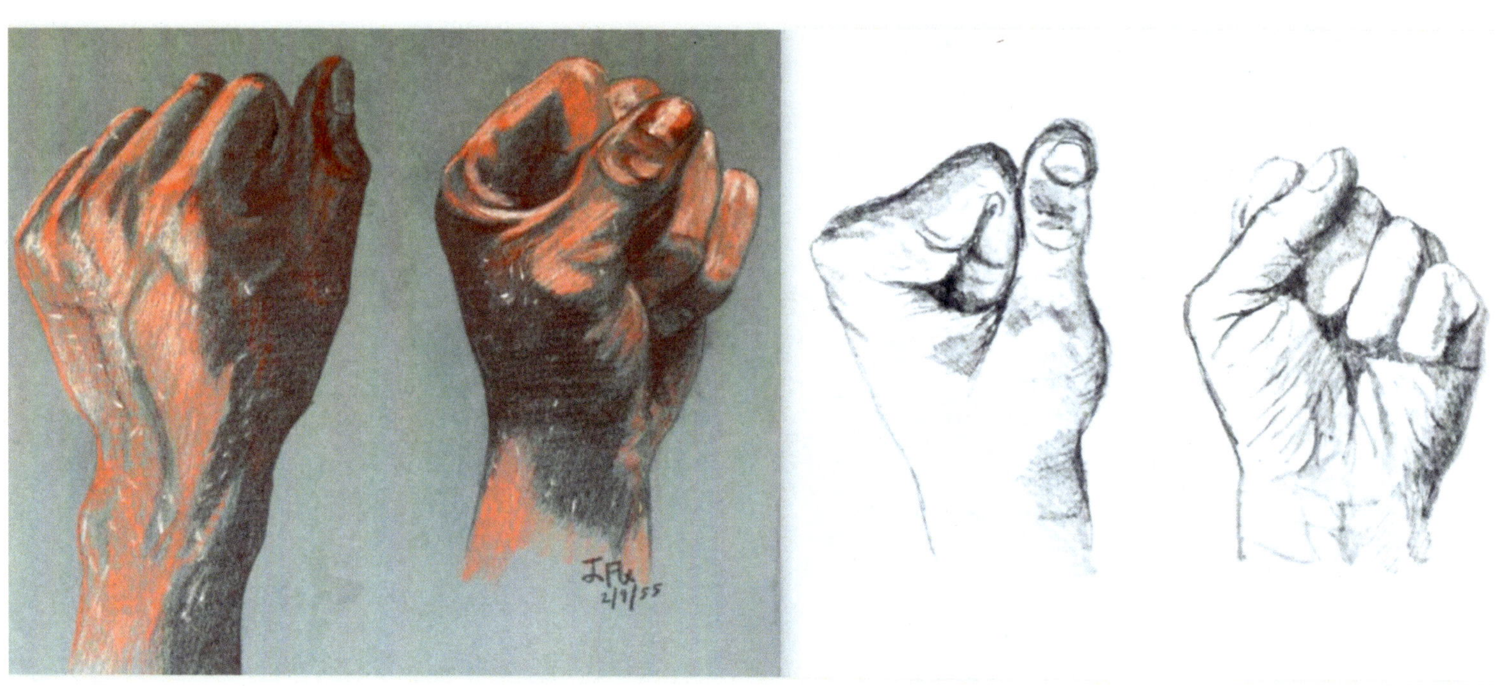

14. We are both right-handed, and drew left fists from similar angles. Mine drawn at university.

Number 15.

15. Some Silver Y *(Autographa gamma)* type moths have crests, but these moths are normally shown in side view. I found 3 front views in 1,828 photos online, but only one was from the same angle as ours.

Number 16.

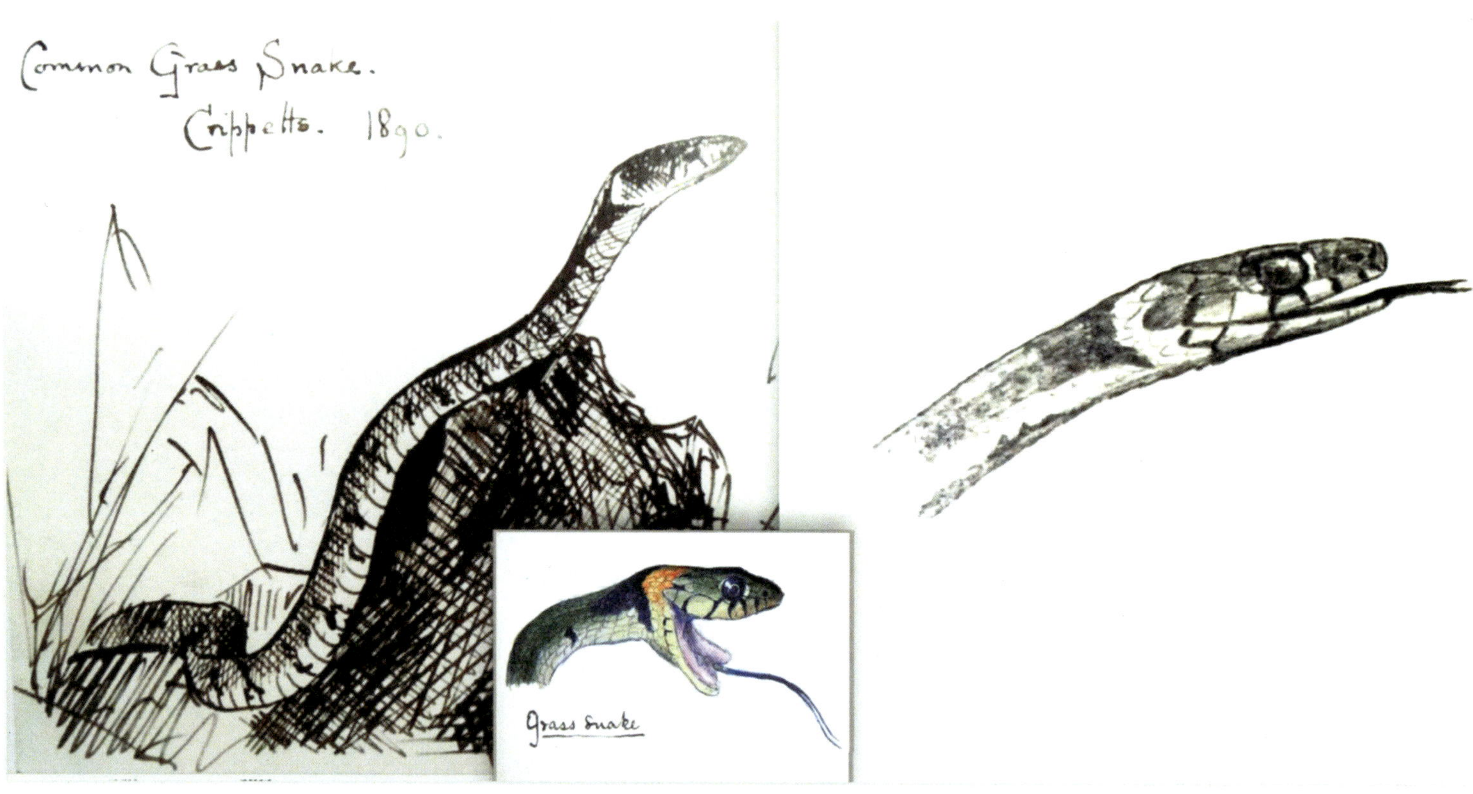

16. The grass snake *(Natrix natrix)* head is from my Zoology dissection notebook; we were not told to draw the head, but I had to record that eye. Wilson comments: "...in that little head and one eye is all the fascinating quickness and subtle gracefulness of all the snakes I have known...". Jane Goodall described a snake's "black expressionless eyes".

Number 17.

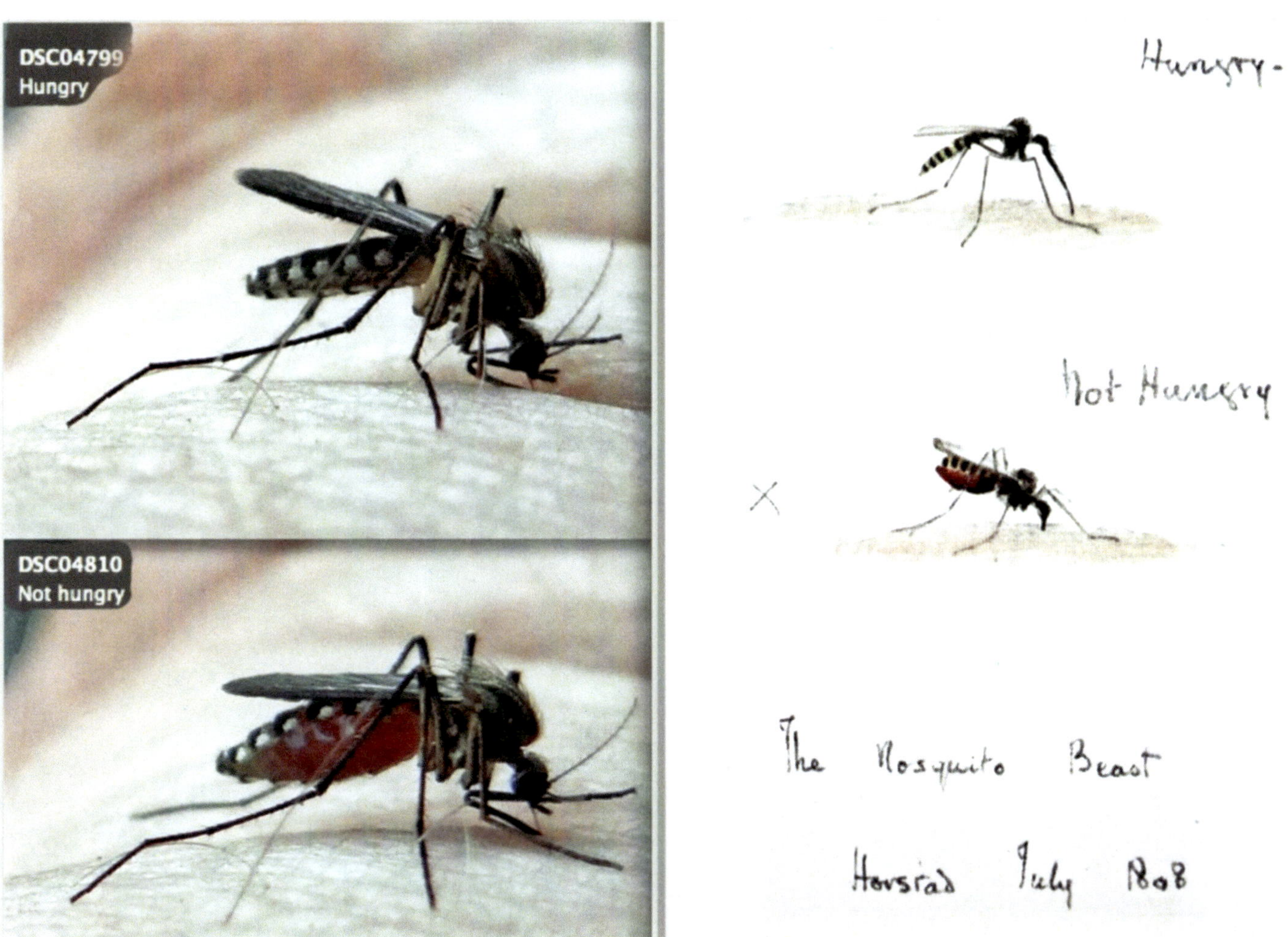

17. We independently labelled mosquitos, from New Zealand and Norway, “Hungry, Not hungry”, rather than empty/full, start/finish, thin/fat.

Number 18.

18. Small tortoiseshell butterfly *(Aglais urtica)* painted from my collection, as a poster for a university talk; and a few from Wilson's childhood collection.

Number 19.

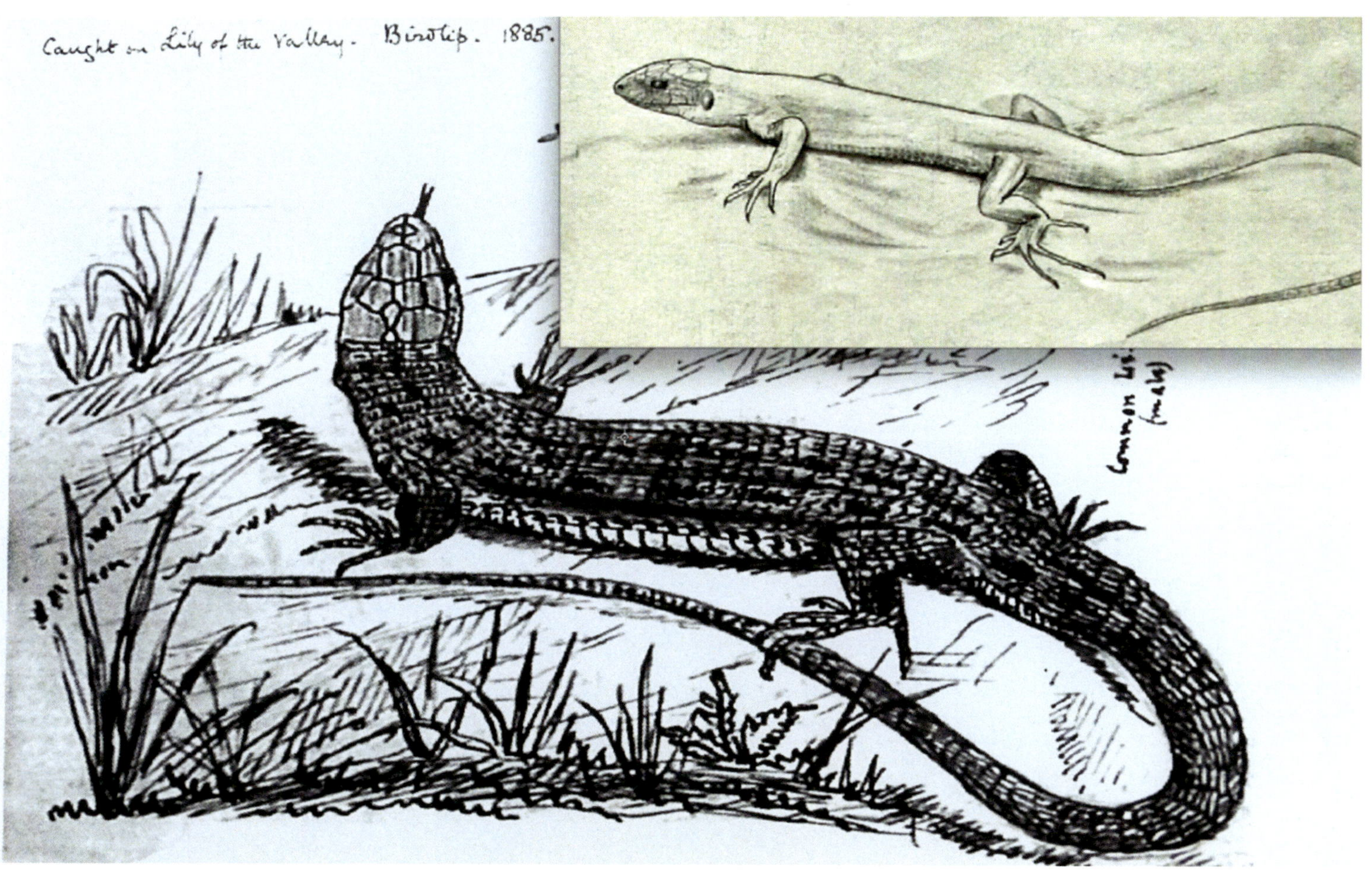

19. Common lizard *(Zootoca vivipara);* their toes and head scales match as they should.

Number 20.

20. Field vole *(Microtus agrestis)*, mine a B&W glass plate photo, showing the open ears that Wilson notes alongside his sketch are needed.

Number 21.

21. Bats, Pipistrelle *(Pipistrellus pipistrellus)* and Noctule *(Noctalus noctula),* similar pose. Reversing mine would make this clearer, but all my images are only aligned for size.

Number 22.

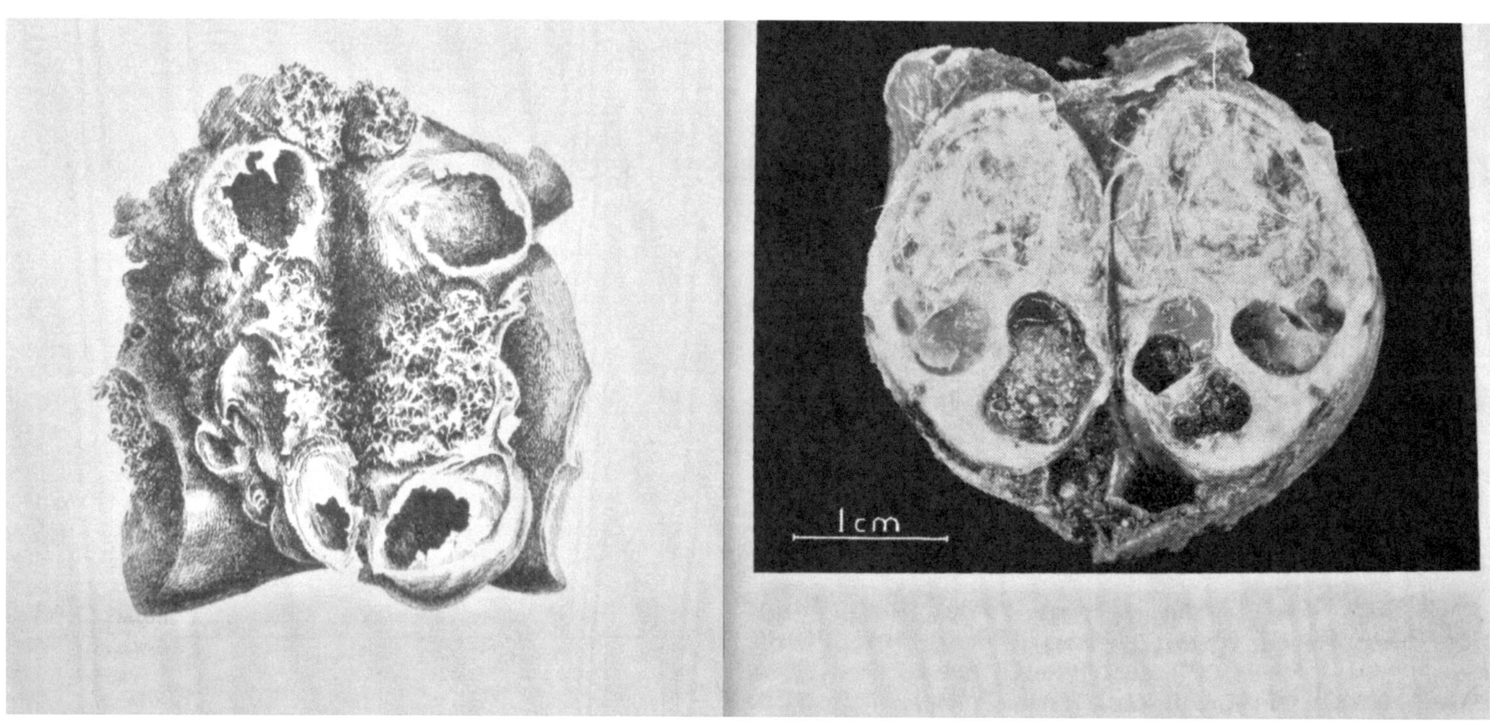

Papillomatous form of carcinoma of the gall-bladder

Hare ovary sectioned to show well-developed teratomata containing hair and other tissues.

22. Cancerous growths. Mine, ovarian cysts in a hare *(Lepus europaeus),* was published in 1965.

Number 23.

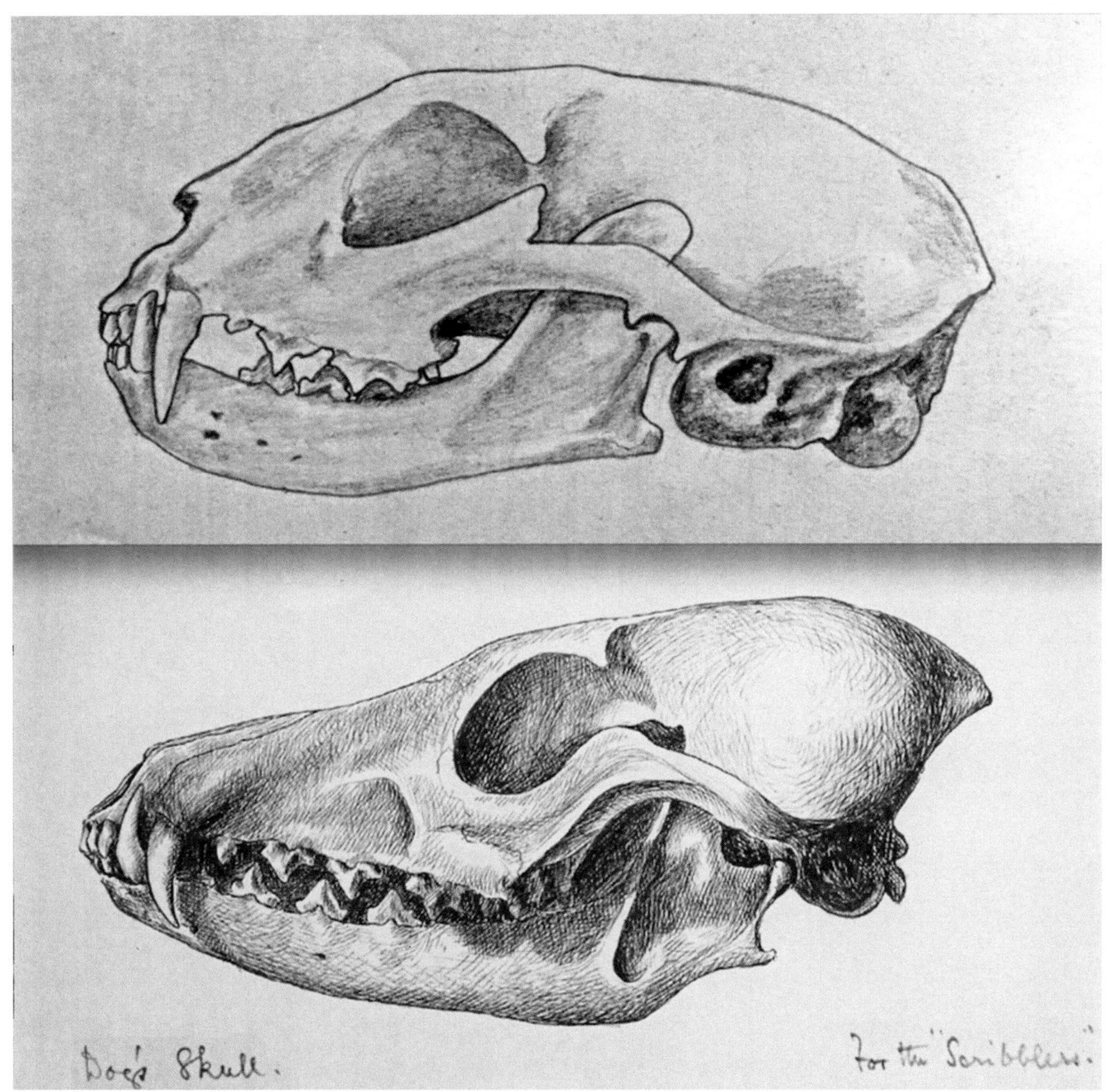

23. Cat and dog skulls; my dog skull was not shaded.

Number 24.

24. Common Porpoise juvenile *(Phocoena phocoena)* I bought in a fish-shop to dissect at university, and Hourglass dolphin *(Lagenorhynchus cruciger)*.

Number 25.

25. Common dolphin *(Delphinus delphis)* and Spotted dolphin *(Stenella frontalis)*.

Number 26.

26. Common seals *(Phoca vitulina)* basking on rocks.

Number 27.

27. Our notebooks; seals at Kerry, Ireland; and Sheep Skerry, Orkney, Scotland.

Number 28.

28. Orcas *(Orcinus orca)*, Antarctica, and a lucky shot showing a youngster in New Zealand.

Number 29.

29. Mountain hare in summer coat. Mine from Finland (1982).

Number 30.

30. Rabbits *(Oryctolagus cuniculus)* in the wild.

Number 31.

31. Wilson had difficulty drawing rabbit ears, but his hare ears were good. Barrett-Hamilton often criticised Wilson's art (done under incredible pressure on the boat to New Zealand) but wrote (in British Mammals): “One could not associate with him without feeling that one had gained something...and no man ever took criticism in better part”.

Number 32.

32. Scottish mountain hares in winter coat. Note faces, ears, and front foot likeness.

Number 33.

33. Stoats *(Mustela erminea)* with the same expressions.

Number 34.

34. Male roe deer *(Capreolus capreolus).*

Number 35.

35. Female roe deer, similar stance.

Number 36.

36. Male Fallow deer *(Dama dama)* looking away.

Number 37.

37. Reindeer *(Rangifer tarandus)* in Norway, and Cairngorm mountains, Scotland.
My half-hidden male in the same pose as Wilson's

Number 38.

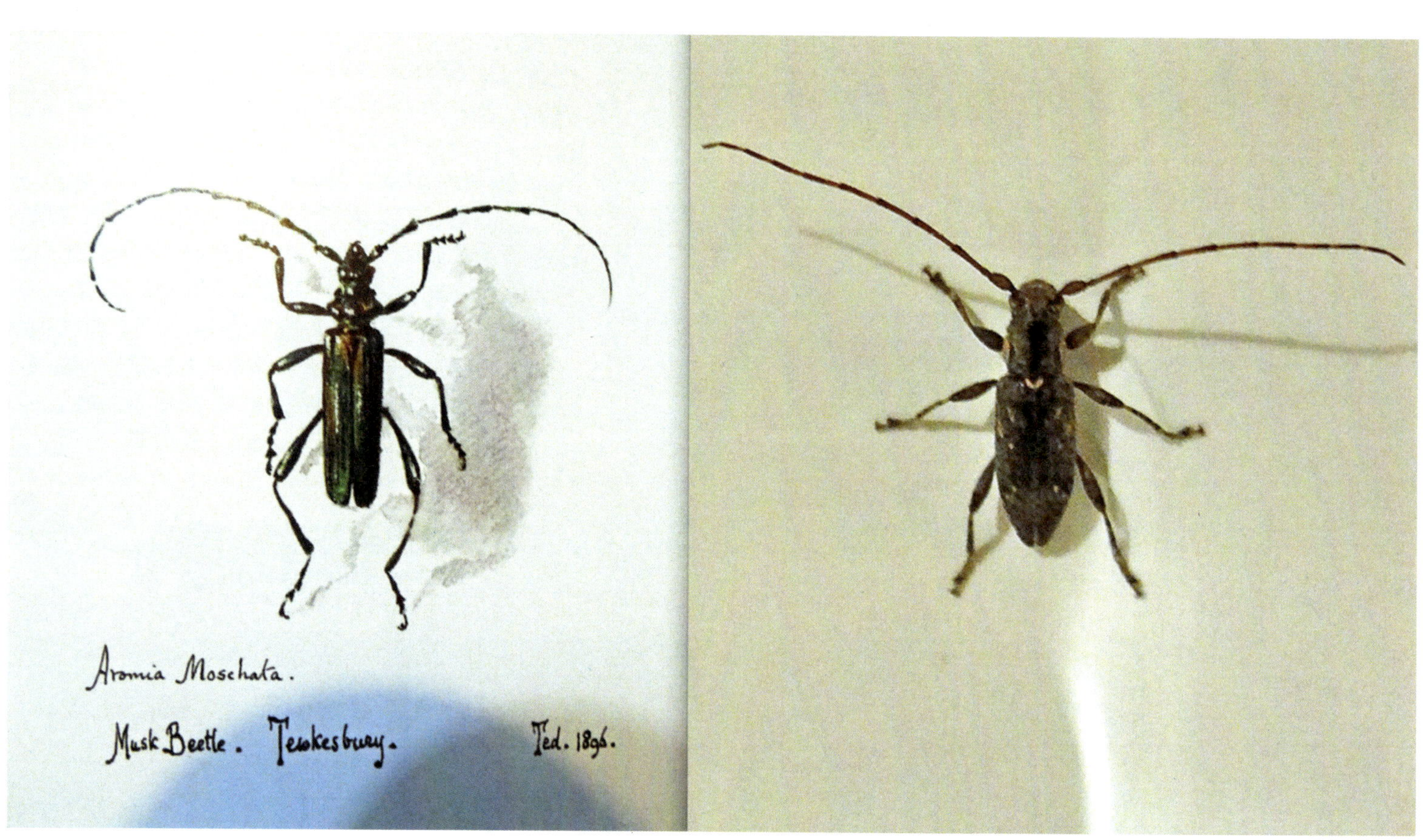

38. Longhorn beetles, showing shadows.

Number 39.

39. Flies. My flies were mating on the clothesline

Number 40.

March. 23. '98.
found all over the gorse.
Crippetts.

40. Seven-spot Ladybirds *(Coccinella septempunctata)*, Scotland 1958, England 1898.

Number 41.

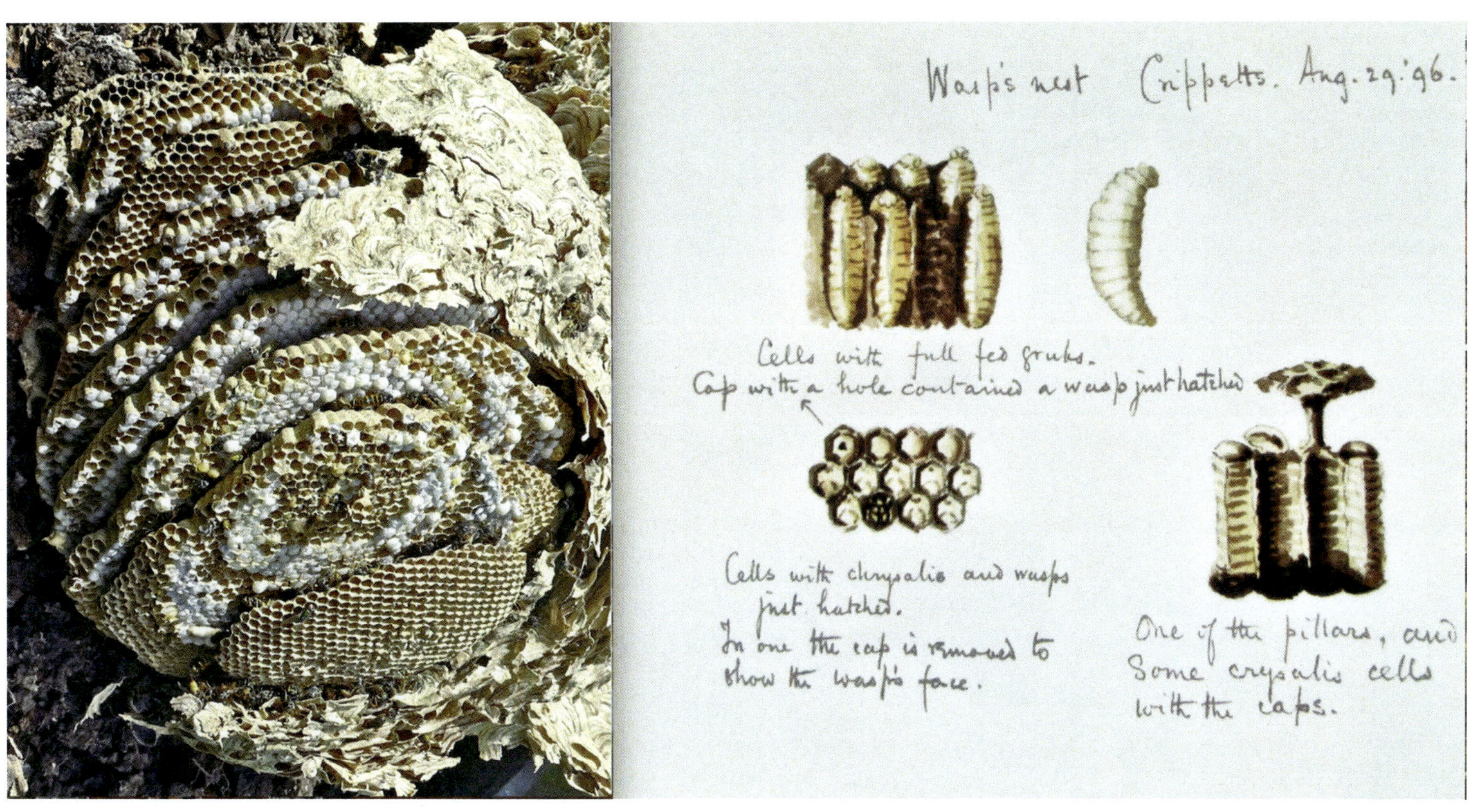

41. Wasp *(Vespula vulgaris)* nests cut open.

Number 42.

42. Common shrews *(Sorex araneus)*, both dead on paths, but this is normal as they are often killed by cats and left uneaten.

Number 43.

43. Amphibia; newts *(Triturus cristatus)* and axolotl *(Ambystoma mexicanum)* in Mexico 2013.
(In England with a group, I saw newts but could not stop; in Mexico I was alone.)

Number 44.

44. Grouse, males showing red and black colour forms.

Number 45.

45. Male Grouse on heather moorland.

Number 46.

46. My university Spiny dogfish *(Squalus acanthias)* for dissection, and Wilson's Leopard seal *(Hydrurga leptonyx)* in the same pose.

Number 47.

47. Arctic *(Sterna paradisaea)* and common terns *(S. hirundo)* are hard to tell apart. Same wing positions.

Number 48.

48. Turkeys *(Meleagris gallopavo)* facing off. New Zealand and England.

Number 49.

49. Brent geese *(Branta bernicla);* mine in Jersey, Channel Islands.

Number 50.

50. Lapwings *(Vanellus vanellus)*.

Number 51.

51. Turnstones *(Arenaria interpres)*, Ireland and Scotland.

Number 52.

52. Redshanks *(Tringa totanus),* Shetland and Ireland.

Number 53.

53. Golden plovers, *(Pluvialis apricaria)* London Zoo, and Pacific Golden plover *(P. fulva)* Cook Islands.

Number 54.

54. Rooks *(Corvus frugilegus)*. Adults with bare faces, same pose.

Number 55.

55. Choughs *(Pyrrhocorax pyrrhocorax),* both on Dingle cliffs, Ireland. Wilson's foreground bird in same stance as mine.

Number 56.

56. One of many wild cats *(Felis silvestris)*
in Aberdeen University deep freeze, killed by gamekeepers.

Number 57.

57. House sparrows *(Passer domesticus),* female feeding young.
Many of Wilson's bird paintings were never finished when Eagle Clarke's book was abandoned by the publisher.

Number 58.

58. Pipit *(Anthus novaeseelandiae)* NZ (2013);
and young Wilson's probable Meadow pipit *(A. pratensis)*.

Number 59.

59. Woodpeckers, Europe *(Dryocopus martius),* and Brazil *(Campephilus melanoleucos)* (2010).

Number 60.

60. Starling *(Sturnus vulgaris)* trios.

Number 61.

61. Whitethroat *(Sylvia communis)*, perched on steep grass stems.

Number 62.

62. Hoopoe *(Upupa epops)* in Bhutan (2007).
Wilson's fast sketch is remarkably identifiable.

Number 63.

63. Oystercatcher *(Haematopus ostralegus)* nests with three eggs.

Number 64.

64. Ptarmigan *(Lagopus muta)*, winter and summer plumage, both in Sweden.

Number 65.

65. Polecat/Ferret *(Mustela putorius/ M. furo)* Britain, and New Zealand 1962.
Note Wilson's skill in recording the colours and sheen needed in his final painting.

Number 66.

66. Knot *(Calidris canutus).*

Number 67.

67. Ringed plovers *(Charadrius hiaticula).*

Number 68.

68. Flying birds; grouse on moorland.

Number 69.

69. Herons *(Ardea cinerea).*

Number 70.

70. Hazel grouse *(Tetrastes bonasia),* Europe, and Spruce grouse *(Canachites canadensis)* the Canadian equivalent, Algonquin Park, 1979.

Number 71.

71. Chaffinches *(Fringilla coelebs)*, on crossed twigs.

Number 72.

72. Redpolls *(Carduelis flammea).*

Number 73.

73. Siskins *(Spinus spinus).* Norway and Scotland.

Number 74.

74. Goldcrests *(Regulus regulus)* top views; mine landed on a ship in the North Sea, 1958.

Number 75.

75. Long-tailed tits *(Aegithalos caudatus)*, road-kills.

Number 76.

76. Crossbills *(Loxia curvirostra).* Mine on migration at Sumburgh Head, Shetland, 2015.

Number 77.

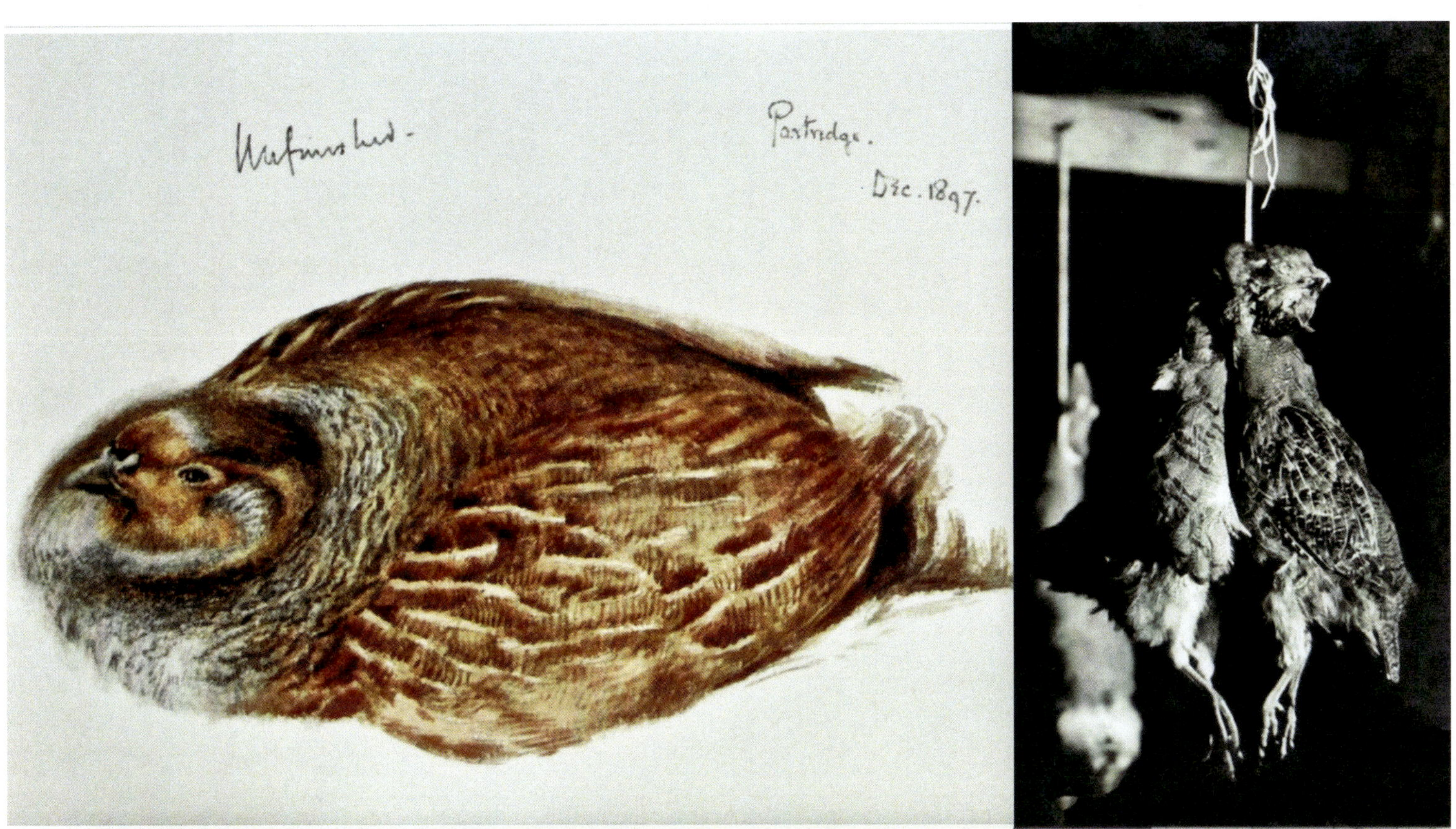

77. Grey partridge *(Perdix perdix)*, and two hanging with a hare *(Lepus europaeus)* in our cellar.

Number 78.

78. Primroses *(Primula vulgaris)* are commonly illustrated.

Number 79.

79. Mountain avens *(Dryas octopetala)* is a less common alpine; these were in the Cairngorms, Scotland, and Norway.

Number 80.

80. Flowers of Bramble/Blackberry *(Rubus fruticosus).*
Wilson drew the torn leaves, as seen.

Number 81.

81. Fruit of Bramble. Both pictures show brown-edged leaves.

Number 82.

82. Gentians *(Gentiana nivalis)*.
Wilson's in Switzerland, mine in German alps (3 May 1977).

Number 83.

83. Moss campion *(Silene acaulis)*, another alpine plant; Norway and Cairngorms.

Number 84.

84. Violets *(Viola tricolor and V. odorata).*

Number 85.

85. Japanese anemone trio *(Anemone hupehensis).*

Number 86.

86. Wilson's Large Wintergreen *(Pyrola rotundifolia)*, and my Intermediate Wintergreen *(P. media)*.

Number 87.

87. Larch *(Larix decidua)* cones.
Mine drawn at school.

Number 88.

88. Red fungi on branches.

Number 89.

89. Toadstools cut in half.

Number 90.

90. Single trees, with dead branches.

Number 91.

91. Honeysuckle *(Lonicera periclymenum)* drawn by my father for me to copy.
(My version lost in Burma/Myanmar, 1941).

Number 92.

Cheltenham College Chapel

92. Church spires.

Number 93.

93. Cathedral interiors; Wells, and Gloucester.

Number 94.

94. Walls and chimneys: "...brickwork in front...and I copied each brick," Wilson noted.

Number 95.

95. Dingle beach, Ireland; both at low tide, into the sun.
The red optical flare in my camera lens matches the red round Wilson's sun.

Number 96.

96. Sunsets over water, looking into the sun.

Number 97.

Iridescent clouds, looking north from Cape Evans, August 9 1911

97. Iridescent clouds

Number 98.

Delbridge Ialands and the slopes of Mt. Erebus. 30 March 1911, 6.30 p.m.

98. Red striped clouds, New Zealand and Antarctica.

Number 99.

Looking South, Mt. Discovery, August 3 1903 1 p.m.

99. Calm seas and sky colours, with mountain peaks. Mine of Mt Taranaki, NZ.

Number 100.

100. Australian Superb Fairy wrens *(Malurus cyaneus),* male and female.

Number 101.

101. Tawny frogmouth *(Podargus strigoides),* also from our short trips to Australia.

Number 102.

102. Ruapehu and Ngauruhoe volcanoes, New Zealand.

Number 103.

103. Maori meeting houses, called Wharenui, New Zealand.

Number 104.

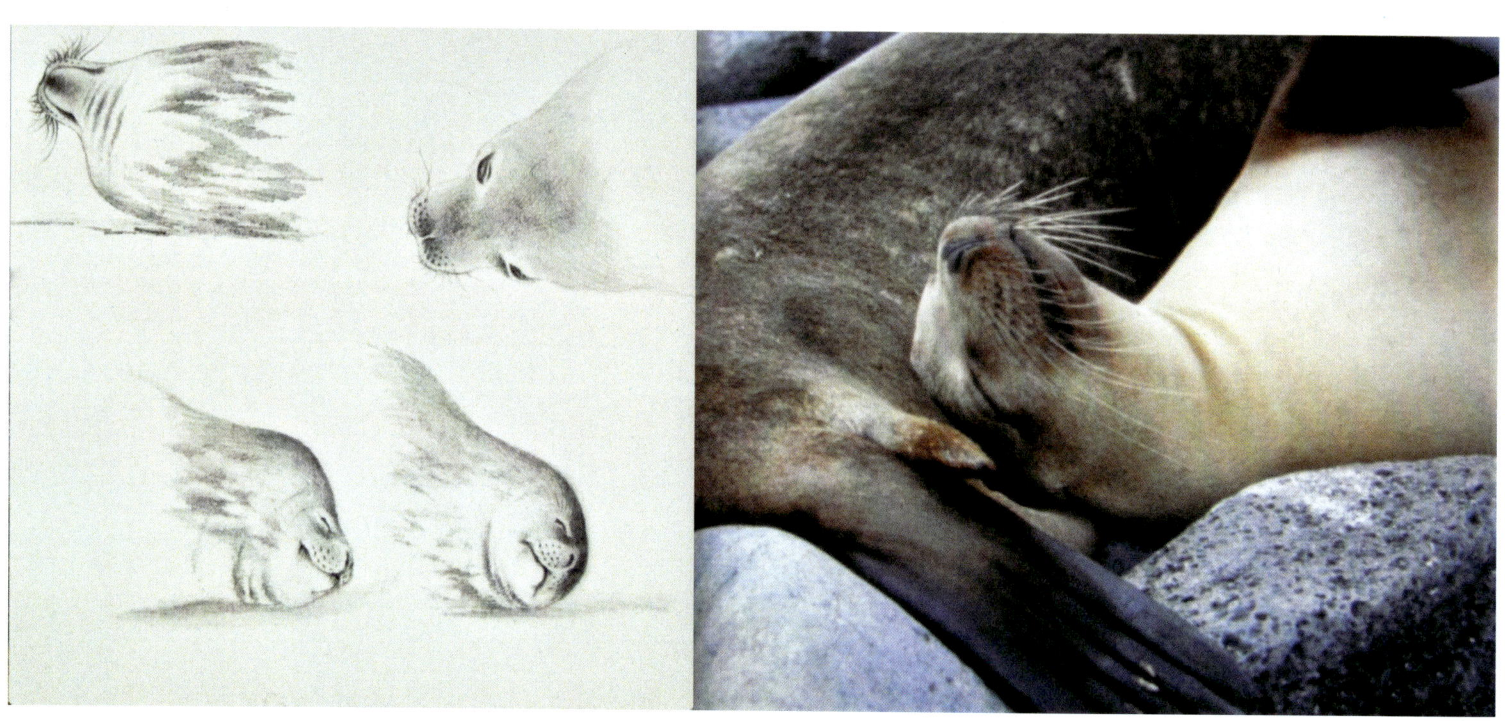

104. Happy female sea lions *(Phocarctos hookeri).*

Number 105.

105. Sea lion males, upright stance.

Number 106.

'Berg off cape Evans, April 23.11. Last day of the sun'

106. Sea caves formed in ice in Antarctica, and rock at Archway Islands, New Zealand.

Number 107.

107. Dippers *(Cinclus cinclus)* UK above and below water;
(C. pallasii) in Bhutan (2007) about to dive.

Number 108.

108. Coal tit *(Parus ater)*.

Number 109.

109. Black-headed gulls *(Chroicocephalus ridibundus)* in winter.

Number 110.

110. Great skuas *(Stercorarius skua)* on Noss, Shetland 1958 defending a nest-site; and in Antarctica at bait. Wilson noted: "...the old birds kept up a continual attack on us when we were anywhere near their young..."

Number 111.

Cat stealing night watchman's supper

111. Cat expressions.

Number 112.

112. Sun halo, Antarctica; and Wellington, New Zealand.

Number 113.

113. Wandering albatross *(Diomedea exulans).* Mine from Cook Strait ferry, 1960s.

Number 114.

114. Composite of my albatross photographs arranged to match Wilson's drawing.

Number 115.

115. Albatross; Buller's *(Thalassarche bulleri),* and Black-browed *(T. melanophris)* head.

Number 116.

116. Wilson's penguin, and Razorbill *(Alca torda)* swimming under water.

Number 117.

117. New Zealand white-eyes/Silvereye *(Zosterops lateralis)*.

Number 118.

118. White-fronted terns *(Sterna striata)*, New Zealand.

Number 119.

119. Heads at university (my only cartoon of any of my lecturers).

Number 120.

120. Pukeko (NZ Purple gallinule/Australian swamphen, *Porphyrio melanotus*).
Left foot pointing to right foot toes.

Number 121.

121. Lapwing wing-shape drawings.

Number 122.

122. Mice.

Number 123.

123. Volcanoes above the clouds.

Number 124.

124. Mist on hills. Wilson's (on right) is Horstad Fjeld, Norway, at midnight.

Number 125.

Spiral Cirrostratus, August 11 1911

125. Spiral cloud formations, Antarctica and New Zealand.

Number 126.

126. NZ Falcon *(Falco novaeseelandiae)*.

Number 127.

127. Fairy prions *(Pachyptila spp.)*.

Number 128.

128. Cape Petrel/Cape pigeon *(Daption capense).*

Number 129.

Frigata ariel wilsoni, a type subspecies of Frigatebird collected by Edward Wilson from South Trinidard

129. Frigate birds *(Fregata spp.).* Overlapping pairs.

Number 130.

130. Dogs in snow. Joe, Bismark, and my son's "Cliff".

Number 131.

131. Goosander *(Mergus merganser)* pairs, Wilson's painting, and on the Dee, Scotland.

Number 132.

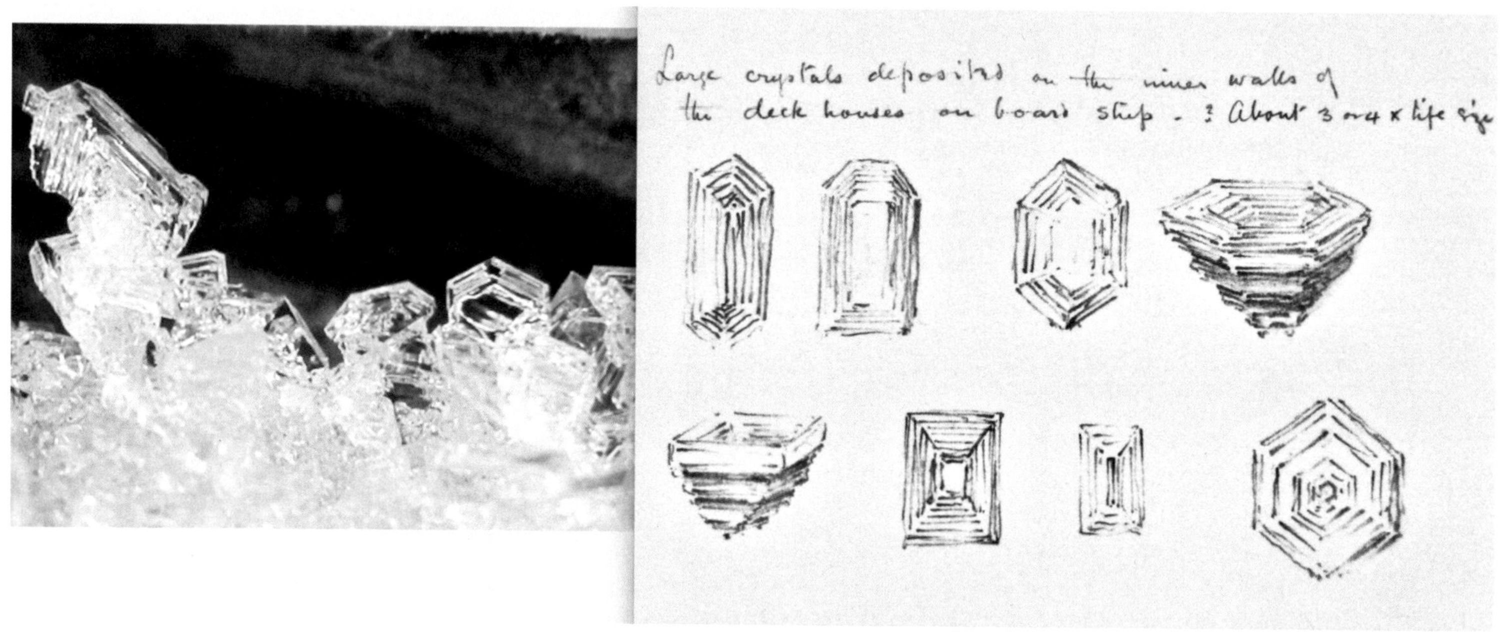

132. Ice crystals of similar shapes, some in the same order. See discussion in text.

Number 133.

133. Tubularia (marine hydroid).

Number 134.

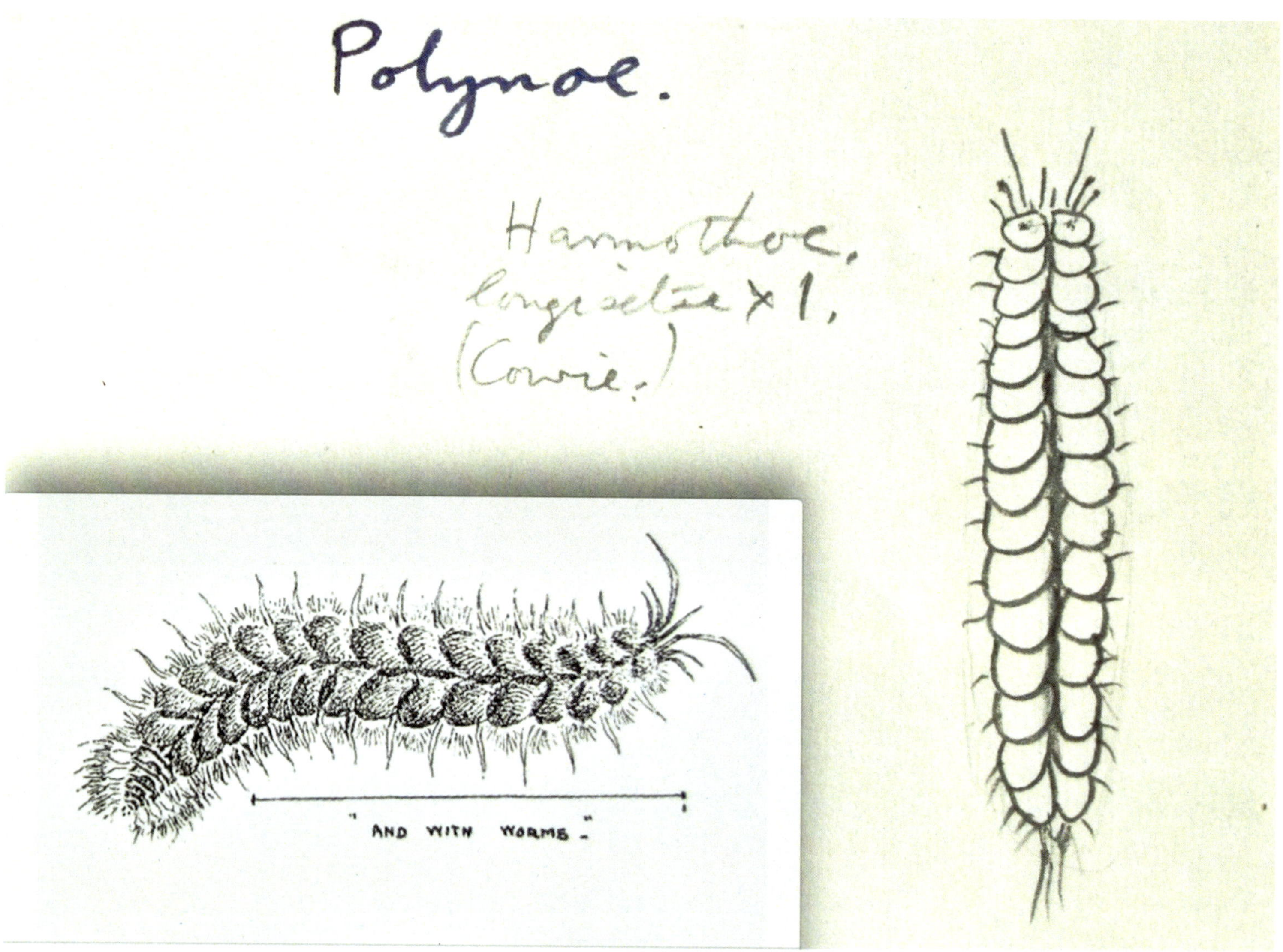

134. Polynoe (marine scale worms, with 15 pairs of scales on their backs).

Number 135.

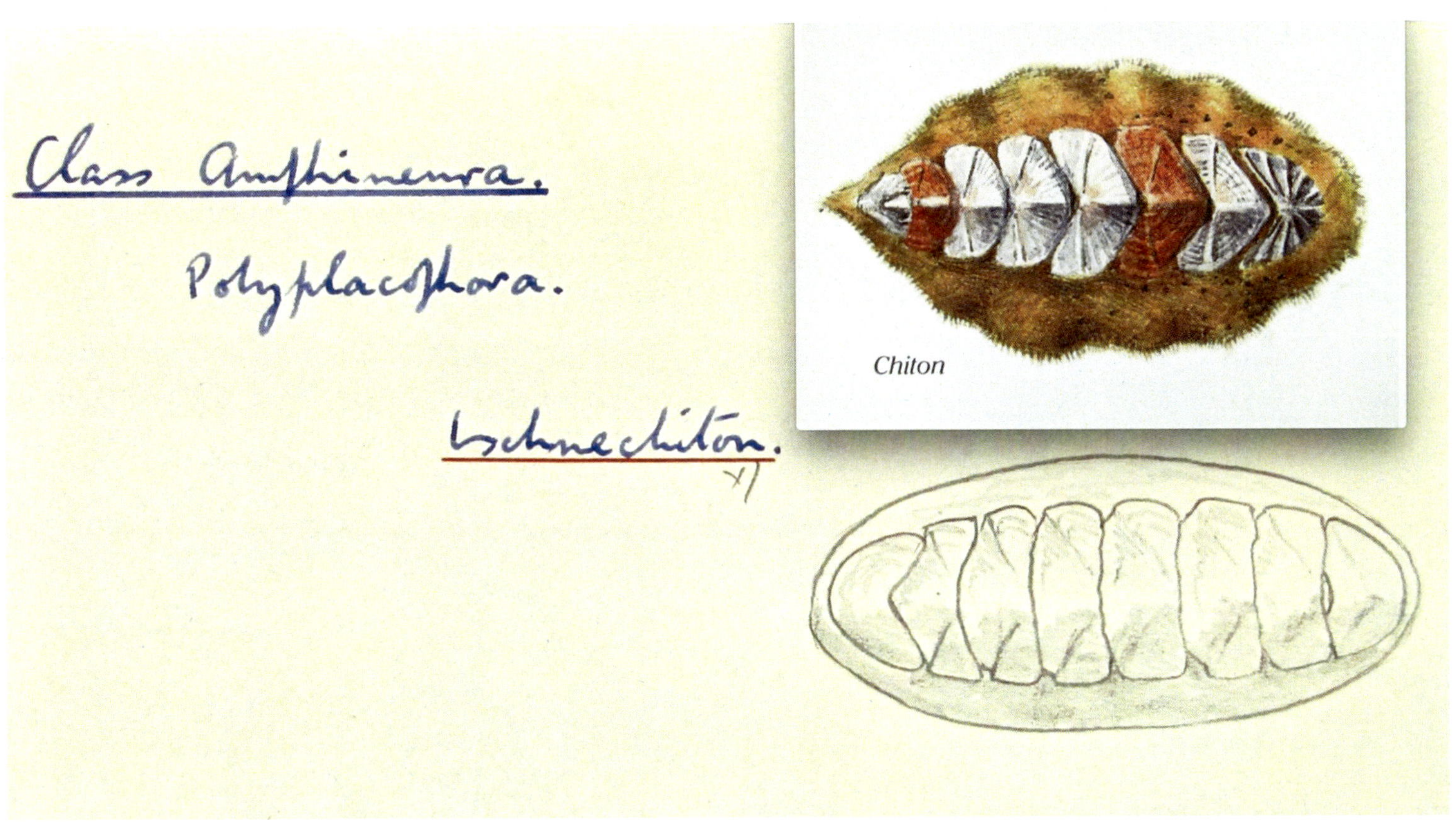

135. Chitons (marine molluscs, with eight scales on back).

Number 136.

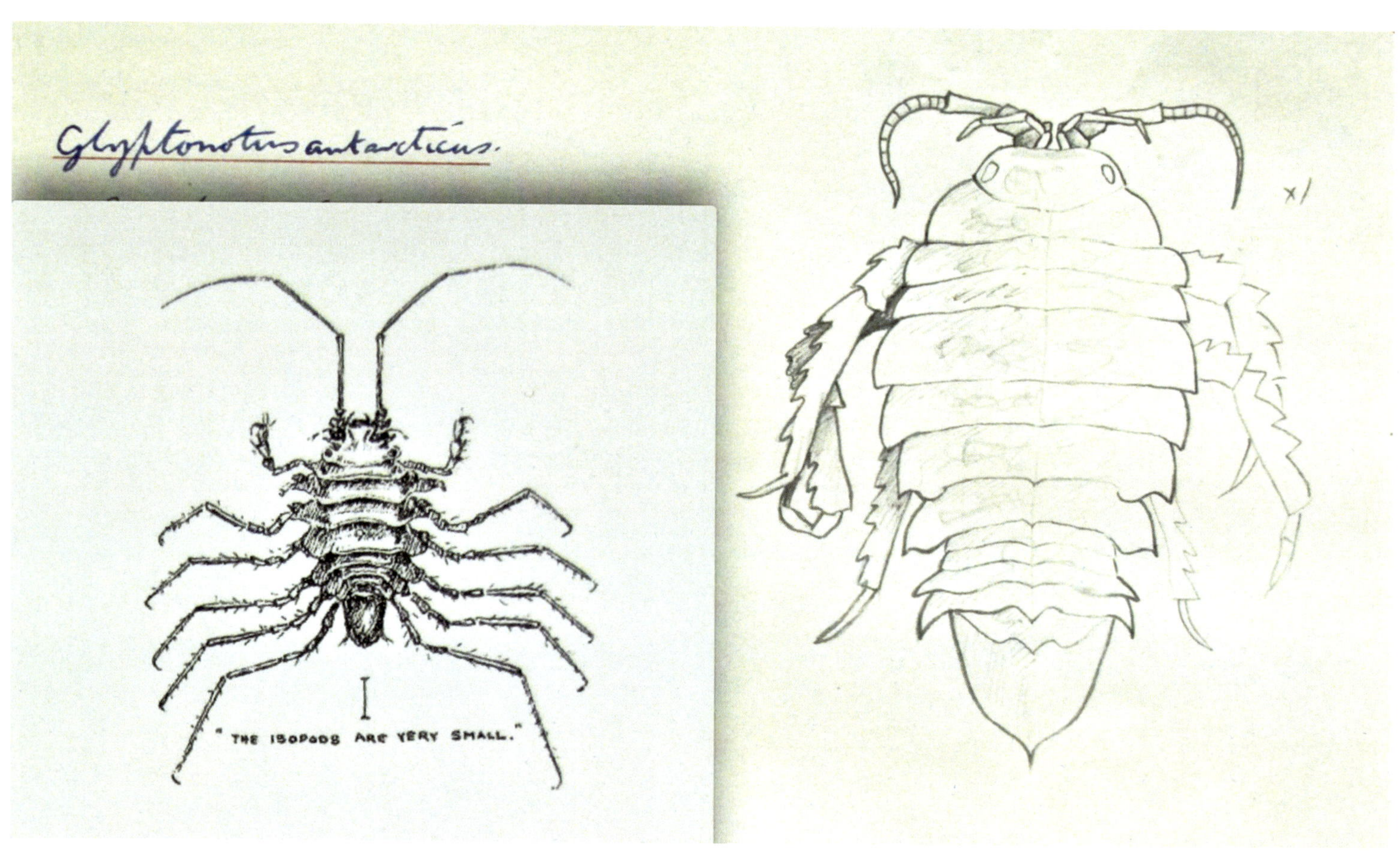

136. Isopods *(slaters).* Wilson drew one of the smallest species *(Munna armoricana),* mine is one of the largest.

Number 137.

137. Met equipment in the snow. Photo of me at Cupola Basin, Nelson Lakes National Park, NZ, by Dick Porter, 2 September1964.

Number 138.

138. White terns *(Gygis alba);* Trinidad, and Cook Islands with small flying fish.

Number 139.

139. Hooded crows *(Corvus cornix).* Mine catching eels in Shetland.

Number 140.

140. Frigate birds *(Fregata spp.)*, Galapagos and Trinidad.

Number 141.

141. Wilson's Adelie penguins *(Pygoscelis adeliae)*
are hatched in a diamond pattern very like that shown in the feathers of the Little blue penguin
(Euduptula minor) from New Zealand

Number 142.

142. Common Scoter *(Melanitta nigra)* heads of males showing enlarged bump on beaks, June 1898 Norway; April 1954 Scotland.

Number 143.

143. Duck drawings. Mallard *(Anas platyrhynchos)* and Wigeon *(Mareca penelope).*

Number 144.

144. Wigeon; mine at Peter Scott's wildfowl park, Slimbridge, UK, 1950s.

Number 145.

145. Black guillemots *(Cepphus grylle)* Norway, and mine at Eynhallow, Orkney, 1958.

Number 146.

146. Gannets (*Morus serrator NZ*, and *Morus bassanus UK*).
(I prefer Wilson's working sketch to the final blurred painting - and lost head colour.)

Number 147.

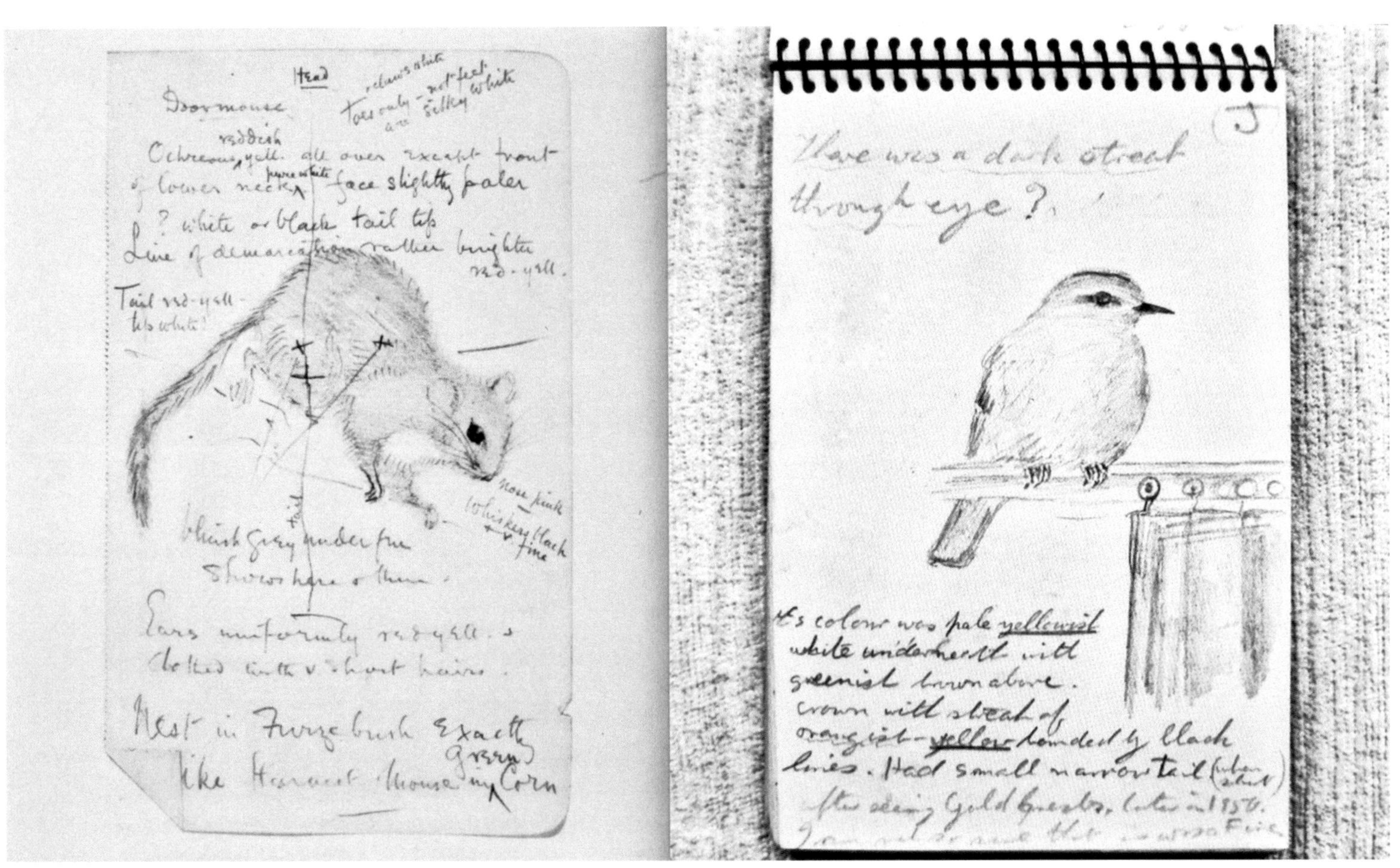

147. Pages from our notebooks.

Number 148.

148. Back views: (hare on back cover of "Lagomorph newsletter" I was editing, 1987-91).

Number 149.

149. Our unconventional diving style. We both used the less common shoulder-width arms, rather than locked thumb/overlapped hand positions (inset). As a diving coach for seven years, I could separate any diver by appearance and style; but not from these photographs.

Number 150.

150. At university we looked alike. I have no idea why I am matching Wilson by wearing a white shirt and funeral black tie, on what seems to have been a bird-watching trip at some beach with a university friend, who sent me this photograph.

Following Edward Wilson into the unknown

Into the unknown

Coincidences

As I read more about Wilson, and the remarkable effect he had on everyone he met, these similarities in pictures became secondary to the overlap in our upbringing, personality, careers, and attitude to nature.

What I consider a "normal" coincidence was that Wilson was born on 23 July 1872 at 6 Montpellier Terrace, Cheltenham, England. On that date in 2019 I would have arrived back in New Zealand from a World Lagomorph meeting in Montpellier, France, if I had not had to cancel my ticket because of the covid19 epidemic. But Wilson had hoped to work in New Zealand on his return from the polar expedition because, as he wrote to his father from New Zealand, "In a century, or less, all or most of this unique fauna and flora will be extinct – they are dying out before one's eyes. I could spend a few years here with advantage on a really classical piece of work". And "He had applied to everyone he could think of in New Zealand to become a government scientist there, or warden...but no post was forthcoming" (Wilson, D.M. and D.B. Elder. 2000). My son has followed this line of research, in the NZ Department of Conservation, protecting kiwi *(Apteryx spp.)*, kokako *(Callaeas wilsoni)*, and other native birds.

My own career was as a government scientist in Ecology Division, NZ Department of Scientific and Industrial Research, working on introduced pests: hares, rabbits, and starlings. I also pursued sidelines on any ecological topic, which were encouraged in those enlightened days. The Division even supported a year's work in East Africa (1967-68), in preparation for which my wife and I learned Swahili, aided by a Methodist missionary couple en route to Kenya: Wilson wrote, "I had hoped to have joined the Universities Mission in East Africa, but this business in my lungs put it out of the question."(Seaver, 1937). Apart from Britain, New Zealand and East Africa appear to be the only countries Wilson wished to work in, and the only places I have worked.

(George Seaver, as early as 1937 in "Edward Wilson: Nature-lover", laments:

> "Nowadays it is evident on all hands that the life of the wild cannot survive the inroads of civilization; and yet there is little doubt that intelligent legislation could minimize the extermination of so many flora and fauna. His [i.e. Wilson's] plea for their preservation in a young colony more than thirty years ago has a pathetic significance today, when it can be paralleled with similar conditions the world over...in an age which sacrifices every natural loveliness to any human convenience").

Personal similarities

Being naturalists from an early age can account for our preference to be alone, or vice versa, and our dislike of team sports: Wilson was "...only just in time to escape the compulsory games act at Cheltenham..." and I kept to the back while our games master picked sides. I knew we leftovers would be sent on a 7-mile cross-country run, which would give me time to look for bird nests or climb trees. Wilson and I both hated parties – W.E. Swinton wrote "He could not stand social occasions and had to have sedatives" (Physicians as explorers. Edward Wilson: Scott's final Antarctic companion,1977). I made a pact with my mother that my sixth birthday party was to be my last. Neither Wilson nor I had any interest in competition; and I was most satisfied when coming second, as in every subject at school. Both of us were following, by chance, Albert Einstein: "Be a loner. That gives you time to wonder. To search for the truth. Have holy curiosity". (Einstein also wrote: "Coincidence is God's way of remaining anonymous.")

Wilson learned art from his parents (both artists), his mother giving him lessons from when he was three years old, keeping his paintings after he was five. His father taught him "the importance of observing from nature itself ... to accurately record it ... from 1886 to 1889 Ted carried out a remarkable set of botanical studies under instruction from his father" (D.M. Wilson and C.J. Wilson, 2004, Edward Wilson's Nature Notebooks). My parents were also good artists; my father was overseas, but sent me drawings to be copied and returned to him. When Wilson was a child "the least thing" was said to make him cry (Wilson and Wilson, above); and I amazed my art class, and teacher, by crying, aged 15, when a design I was working on failed.

But other similarities are harder to explain: as shown in the last photograph we looked like twins, we both won first prizes in diving, using the same, unorthodox style; argued with our fathers about subjects to specialise in at university, resulting in me doing a year of medical courses (Wilson's main subject); both Wilson and I graduated with first class honours in Natural History. We worked with teams investigating declines in grouse numbers, probably including the same famous grouse moors in Glen Esk. In our 20's we taught children in Episcopalian Sunday schools. We estimated time elapsed the same way, counting 100/minute: "Hawk hovered while I counted 248 – 2 ½ minutes", rather than inserting a word (e.g. Mississippi one, Mississippi two, to slow the numbers to seconds as normally recommended). In 5-minute bird counts (Dawson and Bull, 1975), counting to 500 avoided having to look at my watch.

Wilson was frugal, "it was a moral choice about materialism" (Wilson and Elder, 2000). I am too, having been brought up in the Second World War as an Aberdonian: and, to avoid lavishing cash to cite Ogden Nash, No Scotlander squanders.

Career coincidences

Being naturalists, we recorded similar things: arrival and departure dates of migrant birds, and their nesting. Comparing our notebooks for 1905 and 1948, with my dates in brackets: swallows 4 April (7 May), cuckoos 20 April (17 May), swifts 5 May (13 May) – swifts may travel faster because they spend less time in Britain. On nesting dates, Wilson noted a willow warbler nest with 4 eggs (probably still laying as clutch size is 5-7) on 1 May, and one I was watching at Stonehaven reached 4 eggs on 23 May 1948. Wilson was then called to Scotland, where he noted "everything two or three weeks behind Cheltenham, naturally."

On 30 April Wilson "Saw a flock of starlings, 60 to 80, feeding in a grass-field. Seems queer when all of them are nesting." We found the answer was that starling breeding is limited by holes to nest in, and there is a large non-breeding population waiting for a vacancy (Flux & Flux, 2015).

We both collected dead animals on roads, and their fleas; insects, especially butterflies; frogs, fish, and deserted young animals to rear at home; and feathers, but stopped collecting eggs (only one from a clutch) in our late teens, and donated our collections to museums.

Wilson found Geometrid moths in Europe "all marked to suit birch and Alder bark" but New Zealand ones can be variable: *Declana floccosa* just prevents birds developing a "search image" to recognise it - every moth is different:

Declana floccosa, moth varieties, all from the same location (Flux & Flux 2019

However, I must emphasize that, despite these similarities, there is one enormous difference between us: in his short life of 39 years he drove himself to produce a staggering volume of superb work under extraordinary harsh conditions. His search for emperor penguin eggs, that disproved the theory that feathers evolved from reptile scales, is described by Cherry-Garrard (1965), as "The worst Journey in the World". (Remarkably, that theory continues to attract support although it is "complete hogwash", according to Gunter Bechly 2023 "Fossil Friday: A Dinosaur Feather and an Overhyped New Study on the Origin of Feathers" who compares a real dinosaur feather in 100million-year-old Burmese amber.) "All the way from Cardiff to Cape Town he was working at full stretch; when not stoking, trimming, bird-collecting, skinning and sketching, to complete his...coloured grouse-skin plates and in forty-eight hours had seven hours sleep...he worked standing up because he fell asleep if he sat down." (Huxley, 1977).
I live a happy, lazy life (apart from a few years when watching hares all night and starlings by day, an 87 hour week; but that's not "work" to an ecologist). Daily bouts of euphoria are normal, which Wilson (in Seaver, 1937) recorded as unusual in his own life:

> "It sounds but little to make an hour of bliss, but it was just heavenly for an hour to feel perfect health, perfect comfort, perfect life. I could feel nothing wrong with myself."

Wilson remained a Christian, whereas I gave up being Episcopalian then Presbyterian for atheism, now pantheism, where we seem to meet. Perhaps we follow the Gospel of Thomas (Gnostic text, 60-250 A.D.), where Jesus is recorded to have said that the "Kingdom of the Father is spread out upon the earth, and men do not see it".
Wilson wrote:

> "...God made nothing dead. There is only less life in a stone than in a bud, and both have a life of their own...".

George Seaver, in "The Faith of Edward Wilson" (1948) explains that "The pantheistic element in his faith was correlative to the theistic. He saw no opposition between them."
In Chinese philosophy I was probably closest to Daoism (Taoism) – nature comes first; Wilson to Confucianism – people and nature together. Yet there is little difference, as Tao Te Ching writes:

> "The virtue of the universe is wholeness, It regards all things as equal."

Replying to a question on his views on "super-naturalism", Wilson wrote that it was

> "a collection of ideas which have helped me, and may possibly help others; although they might make some people think that I was a pantheist and considered the universe as identical with God. I don't: I believe in a personal God who is also omnipresent."

But he had little time for religious protocol:

> "I cannot bring myself to go to church, when the whole creation calls me to worship God in such infinitely more beautiful and inspiring light and colour and form and sound. Not a single thing out here but suggests love and peace and joy..."

The poet S. T. Coleridge, in "To Nature", agrees:

> "So will I build my altar in the fields,
> And the blue sky my fretted dome shall be..."

Seaver (1948) quotes Wilson's comments about an evening service in the church he attended in Paddington:

> "I cursed and swore more during that sermon than I have at anything for an age. The man was a blithering idiot. How I hate him."

Wilson might have had difficulty with New Zealand's "Holy Trinity": rugby, racing, and beer (all of which I now try to avoid). He was teetotal, and considered sport a "relic of barbarity, certain to die out in time in any civilised nation" (Huxley, 1977).

Supernatural

We have both experienced things verging on the supernatural. Wilson in Switzerland awoke in tears from a dream, "a thing so strange that I must write the reason of it while I remember it, so clearly: I could weep still for the bitter sorrow". His detailed account of mobs of people destroying all the flowers in Crippetts runs to over 400 words, ending "Perhaps they have cut down Crippetts Wood. Oh my God, I hope not." Seaver records: "A corner of the Crippitts woods was destroyed for a building site a few years later." I cried in New Caledonia at the time my son in New Zealand was told he had terminal cancer and decided to end his own life; but a more inexplicable event was my wife's strange question on my return home (in the North Island) on Saturday after a week working alone in the South Island: "What happened on Thursday at 11 am? I had a dreadful feeling" she said. It was the day, and time, on 25 May 1965 that three deer hunters fired 10 shots at the hut I was watching hares from, luckily without hitting me. She was Scots, a race often attributed with "second sight", but this was the only occasion.

Captain Scott's wife also recorded premonitions in her diary: “I was very taken up with you all evening. Something odd happened to the clocks and watches between nine and ten p.m.” (Edgar Evans had died, and the polar party started across the Barrier.) and on 7th March her young son Peter said: "Amundsen and Daddy both got to the Pole. Daddy isn't working now.” That day Wilson distributed opium, Oates had to stop, and Scott wrote: “What we or he will do, God knows.” (Huxley, 1977).

Such telepathic transfer of severe emotion or pain travelled 3000 miles across America between Bernard Beitman and his dying father, and there are records from Europe to America. Because Carl Jung's word for it, telepathy, is now used for thought transfer, Beitman coined “Simulpathity”: this, with “human GPS [global positioning system] stories illustrate human capacities yet to be recognized by modern science.”

Unfortunately, Beitman's “GPS stories” involve two forms: the ability to dream or find by chance something of value for yourself, and an ability to locate geographically a family member in distress.

There is also the problem of the widely used satellite GPS, and a scientifically recognised ability of humans to know the direction of home, discovered by Robin Baker in 1980 (Science, 210) and supported by Kirschvink *et al.*(Bioelectromagnetics, 1992) who found magnetite in human brains – the same chemical used by other homing species. Marc Abrahams (2023) “Spacey superpowers” (New Scientist, No. 3435) requested examples from readers with unusual awareness of space, time, or direction. One replied, “I have always been aware of where I was in relation to compass headings. On an ocean liner en route from Australia to New Zealand, I woke distressed in our inside cabin because we were going the 'wrong way'. My husband was not aware of a problem...” (The ship had turned back to search for a missing crew-man.) This ability may have helped Polynesians navigate; but they also used sun, stars, wind, waves, tides, and birds. Mau Piailug, one of the last with this knowledge, led a canoe team 2,750 miles from Hawaii to Tahiti, announcing after 31 days that they would arrive next day; they did (Winchester, 2023, “Knowing what we Know”).

Beauty and Religion

Wilson wrote: "A happy life is not built up of tours abroad and pleasant holidays, but of little clumps of violets noticed by the roadside, hidden away almost...". I found that the most memorable item from a trip through India was a dragonfly *(Neurobasis chinensis)* sitting on a log emitting brilliant green flashes from its hind wings; and in the Amazon it was the flight of a dozen Long-nosed bats *(Rhynchonycteris naso)*, landing upside down in sequence one above the other on a branch to resemble a strip of bark. Peter Fleming describes them in "Brazilian Adventure" (1933) as

> "...the smallest and flimsiest creatures imaginable, so grey, so silent, so altogether negative, that you could hardly be certain that you had seen them at all. They existed only in flight."

Hence I agree with Wilson: "Things of beauty give me the most intense pleasure, which lasts a long time and can be recalled at will for days, months, sometimes years. There is something in it we don't in the least understand."

One major historical influence for Wilson was clearly the life of St. Francis of Assisi, the Italian mystic and preacher who founded the Franciscan order of monks. According to George Seaver "The faith of Edward Wilson of the Antarctic" (1948), there was no one "on whom his thoughts more often dwelt: first, because he seemed to reflect humanly all that was most lovely in the natural world; then, because he seemed to represent most naturally all that was best and truest in mankind; lastly, because his touch upon his own life he felt to be sacred and almost personal"; and that Frank Debenham had remarked on "the striking resemblance in facial expression between Wilson's delineation of St. Francis and the artist (who was himself unconscious of it)". Debenham adds:

> "...had we of the Expedition known of his regard for St Francis, we might have understood better, not so much the things he would do, but the things he did not do – refused to do, in fact... Here was an athlete who ... would scorn competition or a prize."

Seaver (1948) quotes Wilson's admiration for St Francis at length, from a letter to his future wife:

> "I don't know if you have ever been bitten by that man's character as I have been...I admire the man more than anyone I ever heard of, and that's a thing no one can do without trying to follow him. I despair sometimes of ever seeing my way to it, yet I

> always feel that the method and the opportunity will turn up when it's time. Above all things, I admire in St. Francis his broad-minded happiness in everything that savoured of beauty, whether it was birds, flowers, poverty, or sickness, or any other odd thing that came his way, and turned his mind to love and praise and sympathy..."

Wilson's access to the work of St Francis probably included "The Canticle of Brother Sun" :

> "Be praised, my Lord, for our sister Mother Earth,
> Who nourishes and watches us
> While bringing forth abundant fruits with coloured flowers
> And herbs."

> "If you have men who will exclude any of God's creatures from the shelter of compassion and pity, you will have men who will deal likewise with their fellow men."

Rene Dubos, in "So Human an Animal" (1970) recounts that Lynn White, an American historian, in 1966 "pleaded for a new attitude towards man's nature and destiny. He saw as the only hope for the world's salvation the profoundly religious sense that the thirteenth-century Franciscans had for the spiritual and physical interdependence of all parts of nature. Scientists, and especially ecologists, he urged, should take as their patron Saint Francis of Assisi (1182-1226)."

Hence the writings of St Francis are relevant to both of us:

> "Holy obedience confounds all bodily and fleshly desires and keeps the body mortified to the obedience of the spirit and to the obedience of one's brother and makes a man subject to all the men of this world and not to men alone, but also to all beasts and wild animals, so that they may do with him whatsoever they will, in so far as it may be granted to them from above by the Lord."

Coincidences

In his diary for 3 December 1910 describing a storm at sea, Wilson wrote: "Just about the time when things looked their very worst... there came out a most perfect and brilliant rainbow for about half a minute or less, and then suddenly and completely went out. If ever there was a moment at which such a message was a comfort it was just then - it seemed to remove every shadow of doubt, not only as to the present issue, but as to the final issue of the whole expedition - and from that moment... everything came all right."

Reginald Pound, in his book "Scott of the Antarctic" (1966), is rather disparaging of this: "Wilson chose to see the salvation of the Terra Nova as a miracle, hallowed by his glimpse of 'a most perfect and brilliant rainbow'...Seemingly, it was vouchsafed only to his eyes. No one else mentioned it. He was a deeply religious man."

Gull crossing a brilliant rainbow

Later, Bowers was also struck by a curious coincidence: "...everything fitted in to place us on the sea-ice during the only two hours in the whole year that we could possibly have been in such a position. Let those who believe in coincidences go on believing. Nobody will ever convince me that there was not something more." And Wilson agreed: "...what looked so like a series of petty mistakes and accidents was a beautiful pre-arranged plan in which each of us took exactly the moves and no others that an Almighty hand intended each of us to take..." (Huxley, 1977.)

Afterlife

Such responses to coincidences seem a reflection of Wilson's beliefs:

> "Our knowledge of what can happen is very limited, and we understand precious little of what everyone is bound to allow does happen.... I have myself no doubt that the dead are with us somehow, I believe even that they can influence our lives....".

On leaving Ory, his wife, as the Terra Nova sailed from New Zealand on 29 November 1910, his diary records her "waving happily, a goodbye that will be with me till I see her again in this world or the next..." (Wilson, 1972). And in his last letter:

> "My only regret is leaving you to struggle through your life alone, but I may be coming to you by a quicker way". (Seaver 1936).

To farewell his parents: "I have had a very happy life and I look forward to a very happy life hereafter when we shall all be together again." (Huxley, 1977).

And possibly based on his summary of Jesus' conversation with Nicodemus (Seaver 1936):

> "Ours is a double life, distinct though united.
> Our body lives and our spirit lives: each must be born".

(In contrast, Scott's wife, who was told of his death on the ship taking her to New Zealand to meet him, was unaware a ship's officer had been appointed to watch "...in case she tried to throw herself overboard; and had she believed in an after-life, she would have done so. 'But I am afraid my Con [Falcon Scott] has gone altogether, except in the great stirring influence he must have left on everyone who had knowledge of him." - Huxley, 1977).

Many writers, artists, and especially poets, support an afterlife or spiritual companion.
As David Suzuki (1997) puts it in his classic overview "The Sacred Balance: Rediscovering our place in nature,"

"Since poetry began, poets and songwriters have been fighting the mind/body dichotomy, singing their sense of the world, of the body and spirit moving together through the world eternally".

Kurt Vonnegut: "I sometimes wondered what the use of any of the arts was. The best thing I could come up with was what I call the canary in the coal mine theory...that artists are useful to society because they are so sensitive...They keel over like canaries in poisoned coal mines long before more robust types realize that there is any danger whatsoever." (Chicago Tribune Magazine. 22 June 1969).

Lewis Thomas, in "The Medusa and the Snail"(1979) writes:

"We must rely on our scientists to help us find the way through the near distance, but for the longer stretch of the future we are dependent on the poets. We should learn to question them more closely, and listen more carefully. A poet is, after all, a sort of scientist, but engaged in a qualitative science in which nothing is measurable... Musicians and painters listen, and copy down what they hear."

Edgar Allan Poe was one poet who fought back. In "Sonnet - to Science":

"Science! true daughter of Old Time thou art! Who alterest all things with thy peering eyes.
Why preyest thou thus upon the poet's heart, Vulture, whose wings are dull realities?"

And in "Al Aaraaf":

"- ev'n with us the breath Of Science dims the mirror of our joy -"

(I found Poe's poems, by coincidence, in a book I bought for its cover's fantastic - both senses - black, crimson, and yellow-streaked painting: of Dunnottar Castle, Stonehaven, my home town.)

Albert Einstein, as reported by Eugene Mallove (1985) "Einstein's Intoxication With the God of the Cosmos",WashingtonPost: "So Einstein's legacy must include not only his physical theories but his cosmic religion - little known and little shared, until perhaps another age." He challenged the future: "I maintain that the cosmic religious feeling is the strongest and noblest motive for scientific research." And, "In my view, it is the most important function of art and science to awaken this feeling and keep it alive in those who are receptive to it." Central to him, he said, was a "rapturous amazement at the harmony of natural law, which reveals an intelligence of such superiority that, compared with it, all the systematic thinking and acting of human beings is an utterly insignificant reflection".

What Poets say

Thomas Hardy, in “The Haunter”: *“He does not think that I haunt here nightly:*
How shall I let him know That whither his fancy sets him wandering I, too, alertly go?-
...Yes, I companion him to places Only dreamers know, Where the shy hares print long paces, Where the night rooks go;”

Alex MacLean: "*Those we love don't go away. They walk beside us every day. Unseen, unheard, but always near. Still loved, still missed. and very dear.*"

Francoise Hardy (songs): *“Rendez-vous plus tard dans une autre vie, Ailleurs ou ici.”*
And *“L'autre cote du ciel.”*

Robert Browning, in “One Word More”: *“This of verse alone one life allows me*
Other heights in other lives, God willing.”

William Wordsworth's “Ode: Intimations of Immortality from Recollections of Early Childhood” regrets the loss of his early visions but finds comfort *“In the faith that looks through death, In years that bring the philosophic mind.”*

P. B. Shelley, in the last stanza of “The Cloud”:
“I pass through the pores of the ocean and shores...I change, but I cannot die.
I silently laugh at my own cenotaph...I arise and unbuild it again.”

Jalal al-din Rumi:
“Don't mistake me for this human form. The soul is not obscured by forms.”

Rupert Brooke's “The Soldier”: *“And think, this heart, all evil shed away,*
A pulse in the eternal mind, no less
Gives somewhere back the thoughts by England given...”

And in “Dust”: *“We'll ride the air, and shine, and flit,*
Around the places where we died...”

Edgar Allan Poe: "Spirits of the Dead":

"Be silent in that solitude, Which is not loneliness - for then
The spirits of the dead who stood In life before thee, are again
In death around thee - and their will Shall overshadow thee: be still."

J. J. R. Tolkien's poem "I sit beside the fire and think" describes his joy in nature, ending with:

"But all the while I sit and think Of times there were before
I listen for returning feet And voices at the door."

The Bhagavad Gita (Hindu religious text):

"*Those* who *are born must surely die*
And those who are dead must be reborn."

John Denver (songs): *"There is wisdom here, there's so much to learn*
In each brand new day, in our own rebirth..."
And: *"I know that love is seeing all the infinite in one*
In the brotherhood of creatures; who the father, who the son".

H. W. Longfellow: *"Dust thou art, to dust returnest,*
Was not spoken of the soul."

W. B. Yeats: in "Under Ben Bulben"

"Many times man lives and dies Between his two externities,
That of race and that of soul, And ancient Ireland knew it all."

A.C. Swinburne: *"Body and soul are twins: God only knows which is which."*

William Shakespeare: The Comedy of Errors, Act 5, Scene 1:

Duke. *"One of these men is genius* [guardian spirit] *to the other;*
And so of these. Which is the natural man, And which the spirit?"

Even pets have been included, of which St Francis would approve:

"Together again, both person and pet...
The time of their parting is over at last."

Edna Clyne-Rekhy (A Scots artist, whose unpublished poem, based on a Norse legend, was widely circulated in America).

And a poem by Xenophanes tells how Pythagoras interceded to help a dog being beaten, saying:

> *"Stop! don't beat it! For it is the soul of a friend*
> *that I recognised when I heard its voice".*

Archy, a poet unhappy at becoming a cockroach unable to type properly, blamed Pythagorus:

> "dammed be the soul of pythagorus who first
> filled the fates with this notion of transmigration
> of spirits"

Don Marquis (1931), "archy and mehitabel", 1957 Ed, Faber & Faber Ltd., London.

I knew Pythagoras as a mathematician from my school-days, but had no idea he founded a philosophy of reincarnation. I discovered this after my only auditory hallucination on 27 July 2023, when a voice said "Pythagoras" as I awoke, and I checked on the internet. The following day - by coincidence? or as J. W. Dunne foretold? - a cartoon in the newspaper, Munro the crossword cat, by Sharon Murdoch, showed Munro, with a square head and ears in the typical Pythagoras diagram, reading Euclid for Cats. (A similar coincidence followed on 9th January 2024 when I was stung by a wasp for the first time in many years; in next morning's paper Munro was fighting off a stinging insect flying overhead.)

© Sharon Murdoch.
636, used with owner's permission.

Jane Goodall, who studied chimpanzees, was also a poet and "especially drawn to the concepts of karma and reincarnation". Her poem "The Old Wisdom" concludes:

"Yes, my child, go out into the world;
Walk slow and silent, comprehending all,
And by and by your soul, the Universe,
Will know itself: the Eternal I."

(Jane Goodall's and Douglas Abrams' interviews (with Gail Hudson), published as "The Book of Hope: A Survival Guide for Trying Times" 2022, reveals a character "a little like a modern-day Saint Francis of Assisi, surrounded by and protecting all the animals". Her dedication to work seems similar to Wilson's: 300 lectures a year, in different cities around the globe, to spread hope and protect the environment. This is an enormous undertaking for someone born in 1934 - coincidentally as I was - and we met giving talks at a conference in Nairobi on 11 April 1968. She described her chimpanzee study while her husband, Hugo van Lawick, carried their new baby in the background.)

Not all poets agree about an afterlife, of course. Edward Fitzgerald's translation of "The Rubaiyat of Omar Khayyam" shows Omar's doubt:

"Into this Universe, and *why* not knowing,
Nor *whence*, like Water willy-nilly flowing:
And out of it, as Wind along the Waste,
I know not *whither,* willy-nilly blowing."

Therefore he advises us to find heaven here while we can; and although he was a famous mathematician and astronomer, he chooses poetry and music:

"Here with a Loaf of Bread beneath the Bough,
A flask of Wine, a Book of Verse - and Thou
Beside me singing in the Wilderness-
And Wilderness is Paradise enow."

I can find no reference to Wilson's views on this poem, published anonymously by Fitzgerald in 1859. But John Ruskin, whose five volumes Wilson chose as his prize at university, was most impressed:

"2nd September 1863

My dear and very dear Sir,

I do not know in the least who you are, but I do with all my soul pray you to find and translate some more of Omar Khayyam for us: I never did - til this day - read anything so glorious, to my mind as this poem..."

(Published in "Rubaiyat of Omar Khayyam, rendered into English by Edward Fitzgerald, Edited by George F Maine, Illustrated by Robert Stewart Sherriffs." Collins, London, New Edition 1947, revised 1954, Reprint 1971. 224pp.)

One of Wilson's special poems was "Maud" as recorded by Seaver: "Wilson beguiling these long hours of cold wet misery [caught in a blizzard] by re-reading Tennyson's 'Maud', an old favourite of Caius days, the feeling and phrasing of which now touched him with a deeper significance than formerly." This poem fluctuated in popularity, being severely criticised at first for obscurity, then admired for glorifying war *("The blood-red blossom of war with a heart of fire")*, in tune with the times; but opinions changed after World War 1914-1918. I find many passages that still hold that "deeper significance":

"Let knowledge grow from more to more.
But more of reverence in us dwell;
That mind and soul, according well,
May make one music as before..."

There was even a message for me:

"The man of science himself is fonder of glory, and vain,
An eye well-practised in nature, a spirit bounded and poor;
The passionate heart of the poet is whirl'd into folly and vice.
For not to desire or admire, if a man could learn it, were more
Than to walk all day like the sultan of old in a garden of spice."

And advice to follow Tennyson's aim:

"Be mine a philosopher's life in the quiet woodland ways..."

However, philosophers seldom seem to agree with each other: philosophy, although "cultivated by the very best minds which have ever existed over several centuries... nevertheless, not one of its problems is not subject to disagreement..."

(Rene Descartes, 1637, "Discourse on Method and the Meditations"). So I still walk alone, following Descartes' advice, "that it is good to omit doing things which might perhaps bring some profit to those who are living, when one aims to do other things which will be of greater benefit to posterity," by donating more to conserve native forests, than to human welfare.

What recent Philosophers say

An excellent new book by A. C. Grayling (2023) "Philosophy and Life: Exploring the great questions of how to live" gives a very clear review of the contributions made by most of the major philosophers since Socrates initially proposed the key question: What makes life worth living? To quote Grayling:

> "In fact he put the point like this: the *unconsidered* life is not worth living because it is not one's own life, it is at best someone else's idea of what life should be...A life worth living is a life one *owns* because one has *chosen* it deliberately. And 'deliberately' means: on the basis of thinking about it."

As scientists and loners, both Wilson and I seem to have chosen deliberately, and found "lives worth living". After a BSc Honours that included a thesis on trout, and taxonomic collections of freshwater fish and planarians (flatworms), I applied to study salmon for a PhD: but my professor considered I was becoming too specialised. He offered a choice of eider ducks or mountain hares. I wanted bird-watching to remain my hobby, so I chose mountain hares. Hares became and remain my favourite animals and main scientific interest, but ecology encourages sympathetic awareness of the interactions between all life forms, and a desire to protect their environment.

Wilson's character and motivations are well summarised by Elspeth Huxley (1977): "Wilson was that rare phenomenon, a good man whose goodness does not provoke resentment...Perhaps it was his sense of humour that saved him from the taint of sanctimony...Everyone loved Uncle Bill..." As to his motives, "Fame and fortune were temptations of the devil that never, so far as the record goes, tempted him. Few men can have more conscientiously striven to live as a true Christian, neither denying the world nor accepting its false values... He was an ascetic, yet there was something in him of the boisterous undergraduate; he was an artist of delicate and perceptive sensitivity, and a dedicated scientist with a passion for truth: a very complex man. In volunteering for polar exploration he was in part responding to the call of adventure, but in the main seizing the opportunity to learn more of the marvels God had created, and to spread the knowledge of them to his fellow men."

An unusually explicit account of Wilson's effect on other members of the polar team is revealed in Apsley Cherry-Garrard's 'Postscript' added to the 1965 edition of “The Worst Journey in the World”. George Seaver's Foreword: “But if the first part of the 'Postscript' does little to add to our knowledge... the second part of it is sheer inspiration. For it is the expression of his ultimate faith... the poetic feeling, the patriotic sense, the touches of philosophy - they show that despite... life's disillusionment, he kept to the end his early ideals... He sees it triumphantly vindicated in the characters of the men he knew and loved down South.”
“... In such a world, violent, angry and tired, Wilson sets a standard of faith and work. In a world which destroys itself and beauty, desperately and impotently desiring peace, he helps... We have missed him ever since he died. But you must find him: his voice, it is a quiet voice, is for those who listen... by the time he started on our Winter Journey he had reached another plateau... where he was beyond ambition and beyond fear. He had the quiet mind. That feeling he could communicate to others. Such men do occur in history, but they are very rare, and when they do happen they are among the great ones of our race.”

Other aspects of his character, and their relevance to art and the moral codes when he lived, are revealed by a fascinating critique of his ice drawings by Ruth Watson (2011-12), “The art and science of ice: Edward Adrian Wilson's ice crystal drawings”. “Wilson's ice crystal drawings seem remarkably free of predetermined habits. As we look at his drawings we can see Wilson looking, marvelling and trying to understand ice crystals' forms and methods of growth. His drawings direct our attention to the subject and suggest the questions that need to be asked. This is a method without words (notably, in Wilson's work, the words are the notational device, not the drawing) and seems much more modern than we might expect...this art could be said to be as explorative as the expedition itself... Wilson drew because he had to; it was an interior impulse that would likely have flourished in any setting, and it was informed by his own brand of Christianity as well as Ruskinian tenets chiming with those beliefs.”
And, from Stevens, O'Connor, and Robinson (2019), “The connections between art and science in Antarctica: Activating Science*Art”: “... this artistic view deviates from the so-called scientific method, but the field of science is littered with advances built on the intuition of very clever people.”
Leonardo da Vinci had expressed this link four centuries ago, with an ecological insight: “To develop a complete mind; Study the science of art, study the art of science. Learn how to see. Realize that everything connects to everything else”.

Wilson clearly portrayed the beauty and significance of ice formations in Antarctica: I had access only in Scotland, but this image shows how a dark surface (wire) attracts heat to throw off ice, now an important concept driving sea-level rise following climate change.

Wire and ice fences. Lecht, Scotland, 1958

To return to Hamlet. Boris Pasternak's short poem "Hamlet" is based on an intimate knowledge of the play (he translated it into Russian), and ends:

> *"And yet the order of the acts is planned,*
> *The way's end destinate and unconcealed.*
> *Alone. Now is the time of Pharisees.*
> *'To live is not like walking through a field'."*

Pasternak comments (and I find a curious similarity to Wilson's character): "Hamlet is not a drama of a weak character but a drama of duty and self-denial. When appearance and reality are seen to diverge and are separated by a gaping chasm it is of slight importance that the warning as to the world's falsity should come in a supernatural way and Hamlet be summoned to revenge by a ghost. It is of far greater importance that by the merest accident Hamlet should be chosen to sit in judgement on his time and become the servant of a remoter one... to do the will of Him that sent him." (Boris Pasternak "In the Interlude Poems 1945-1960". Translated into English Verse by Henry Kamen, 1962). This translation is remarkably fine poetry, and by coincidence Kamen was born in Burma/Myanmar, as I was.

Coincidences

But I hold in mind Stephen Jay Gould's warning in "The Flamingo's Smile" (1985, p. 119):

> "The human mind delights in finding pattern – so much so that we often mistake coincidence or forced analogy for profound meaning. No other habit of thought lies so deeply within the soul of a small creature trying to make sense of a complex world not constructed for it."

And in "An Urchin in the Storm" (1990, p. 206):

> "Sure we fit. We wouldn't be here if we didn't. But the world wasn't made for us and it will endure without us."

Gould's articles have always enchanted me, and as another coincidence I quote an editorial by Caroline King, Senior Editor, in New Zealand Journal of Zoology (2015), introducing an essay based on 50 years studying starlings, "The fertility clinic: a bird's-eye view of our future".

> "Very few New Zealand scientists can claim to follow in Gould's rather large footsteps, but one who can is John Flux".

I bought Gould's book "Rocks of Ages" soon after it was published, in 2002, but found it was all religious stuff and never read it. (At that stage I was more in tune with Oscar Wilde - "Religion is like a blind man looking in a black room for a black cat that isn't there, and finding it.") Things change. Now I find his separation of science and religion as mutually exclusive disciplines, tolerant, and useful. Here Gould returns to his topic above:

"To anyone... discouraged at the prospect of life as a detail in a vast universe not evidently designed for our presence, I offer two counter-arguments and an item of solace." These are: (1) the intellectual challenge; (2) Socrates' advice to 'know thyself'; and (3) "a wonderful sonnet by Robert Frost". This poem, "Design", describing three white items encountered on a walk, starts:

"I found a dimpled spider, fat and white,
On a white heal-all [flower], *holding up a moth*
Like a white piece of rigid satin cloth -"

Such a peculiar combination Frost thinks, according to Gould, "must record some form of intent; it cannot be accidental. But if intent be truly manifest, then what can we make of our universe – for the scene is evil by any standard of human morality." So Gould relegates it to chance.

However, chance has not finished. I published a photograph of a white orchid flower, on which a camouflaged crab spider (which has an odd-shaped body) was eating a white moth, with its stiff wings dangling exactly as Frost describes - "like a paper kite" (Flux 2016, Illustrate Ecology: Fertiliser to fertiliser. NZ Ecological Society Newsletter). The orchid is probably fertilised by moths, and I conclude "Doubtless the orchid benefits if a certain proportion of its fertilisers are turned into fertiliser."

Interestingly, Wilson had a more ecological view than Gould of the "morality" of nature. After Wilson's long, and morbidly detailed, account of how an abandoned penguin chick was torn to death by a skua, Seaver (1937) records:

> "It is characteristic of Wilson that he should continue so graphic and sympathetic a picture of Nature at its most merciless and savage with the words, 'Not once nor twice, but a thousand times this happens, and the kindness of Nature's seeming cruelty was borne in upon us as we watched its working."

As Hamlet agreed, "... for there is nothing either good or bad, but thinking makes it so." (Hamlet, Act 2, Scene 2, 248.)

Another Antarctic multi-coincidence.

Of Lawrence "Titus" Oates (who walked out of the tent to die, leaving Bowers, Scott and Wilson), Ranulph Fiennes writes: "... his physical sufferings would have died away as nerve endings froze... and the spirit of Titus Oates left Antarctica, bound perhaps for the stables of Gestingthorpe in Essex."

And Fiennes adds: "I find myself much mirrored in him and empathise with many of his reactions to life. We both lost our fathers when we were young and were blessed with loving mothers. We came from old English families without financial worries. We could not pass exams, nor understand basic mathematics. We went to Eton and failed to enter university. We both desperately wanted to be career officers in the same cavalry regiment, the Royal Scots Greys. We saw action in wild countries against our country's enemies but we never made it to promotion beyond captain and ended up in the polar wastes. Like Oates, I always wanted to be boss in my own fiefdom..."

Ranulph Fiennes (2003) "Captain Scott", Hodder & Stoughton, London. This book is well researched, beautifully written by one who can convey the hardship from personal experience, and has excellent illustrations. Reading it I felt far more involved with Antarctic exploration.

The Future

Finally, when Wilson was born in 1872, the world population was 1.5 billion; when I was born in 1934 it had just passed 2 billion, the maximum ecological carrying capacity; it has now exceeded 8 billion. Our 2018 paper, "Deevey's Hare and Haruspex revisited: Why domestication dooms civilisation?" explains why either humanity or the natural environment will survive in this increasingly over-crowded world; but not both. Gould's "standard of human morality" is restricted to our domesticated "civilised" form which, like all domesticated species selected for tameness, has lost the "amoral", aggressive, population regulating mechanisms of earlier societies - and apparently of all other successful wild vertebrates according to Wynne-Edwards (1962) "Animal Dispersion in relation to Social Behaviour". Therefore, unless an acceptable way to curb, and then reduce, our population can be found (which seems highly unlikely) extinction following expansion to the food limit is inevitable. Jared Diamond in "The Worst Mistake in the History of the Human Race" also suggested this change to agriculture was "our greatest blunder" (Discover 1 May, 1999).

(A curiously optimistic book, "How Peace Came to the World", edited by Earl Foell and Richard Nenneman, was published in 1986. Based on over 1300 essays entered in the Peace 2010 contest - to imagine our world 25 years into the future - it concluded: "Nothing threatens us today more than the 50,000 nuclear warheads that stand in a state of near readiness around the globe. Yet in the year 2010 the world is at peace and the threat of nuclear devastation has vanished." Unfortunately the Doomsday Clock tells a different story, from 6 minutes in 1988 and 2010, to 90 seconds in 2024. And Kenneth Boulding's contribution seems the only one to mention population growth as a problem).

Is it just a coincidence that the major environmental problems we face are: rising temperatures and wildfires, loss of soils, an atmosphere with increasing carbon dioxide and methane, and critically polluted rivers and oceans - the four classical elements that the Greek philosopher Empedocles listed in 450 BC: fire, earth, air, and water?

As long ago as 1858 Francis Bacon saw that "we create worlds, we direct and domineer over nature, we will have it that all things *are* as in our folly we think they should be, not as seems fittest to the Divine wisdom, or as they are found to be in fact." He advises us "to approach with humility and veneration to unroll the volume of Creation, to linger and meditate therein...to study it in purity and integrity." Spedding, Ellis, and Heath, Eds., "The Works of Francis Bacon, Vol 5, p.132."

Forty-two years ago (1981) one of New Zealand's best essayists, Dave Witherow, warned:

> "The derisory concessions that are now granted the environmental movement can only help delay a more general awareness of the simple choices that confront us: we can have growth (for a little longer, and at great cost), or we can begin to build a society that will cease to rip apart the very fabric of the Earth."

Jane Goodall (2022): "And by 2050 there will apparently be closer to ten billion of us. If we carry on with business as usual, that spells the end of life on Earth as we know it".

David Attenborough, scientist and TV personality, also warns us clearly:

> "We are a plague on the Earth... Either we limit our population growth or the natural world will do it for us..."

> "… anyone who thinks that you can have infinite growth in a finite environment is either a madman or an economist." "…we have no moral right to exterminate for ever the creatures with which we share this earth."

(The Living Planet: A Portrait of the Earth. Readers Digest augmented and enlarged edition, 2010, Readers Digest, in conjunction with William Collins, and the British Broadcasting Corporation; also Wikipedia.)
Asked to comment on this by LiveScience, Paul Ehrlich, (who wrote "The Population Bomb", Sierra Club, 1969) said: "I completely agree, as does every other scientist who understands the situation". This is backed up by the 20,000 of us who have now signed "World Scientist's Warning to Humanity: A Second Notice" William J. Ripple *et al.*, (2017).

Rob Dunn (2022) in his excellent book, "A natural history of the future: What the laws of biology tell us about the destiny of the human species", reaches a similar conclusion. "We once assumed the story of life was about us. Now we know that the story of life is mostly about microbes [We consist of 30 trillion human cells, and 38 trillion bacteria - Sender *et al.* 2016 "Revised Estimates for the Number of Human and Bacteria cells in the body"] … As the paleontologist Stephen Jay Gould put it in his book "Full House", 'Our planet has always been in the Age of Bacteria'...Once the ants are gone, it will remain the age of bacteria... We are a clumsy giant late to the drama, a character in life's play that doesn't make it to the curtain call."

Epilogue

Most writers insist we must not give up hope; many claiming (for several years now) we have ten years to fix the problem. A good start is Colin Butfield and Jonnie Hughes' 2021 book "Earthshot: How to Save our Planet". The introduction by HRH Prince William sets a high standard, and the five annual prizes of 1m pounds each year for the next ten years is an excellent way to bypass government inaction: "Pessimism and despondency must be turned into optimism and action".

Hannah Ritchie's 2024 book "Not the End of the World: How We Can Be the First Generation to Build a Sustainable Planet" should be a winner, and certainly is a plan that could save much of humanity. (However, at least four apocalyptic problems remain to be solved: congenital ignorance, fanatic religions, selfish greed, and atomic war.) Tim Flannery 2015 "Atmosphere of Hope: Searching for solutions to the climate crisis" has a similar theme: "Younger people need to be given the chance to create a better world for themselves... I want them to know there is hope - that their new-found voice is making a difference..." (Since Flannery wrote that, Greta Thunberg's leadership gave me hope, and I walked on all local 'School Strike for Climate Change' marches; but few adults joined us. Most people do not care.) Chris Packham, a UK broadcaster extending David Attenborough's reign, who recently legally challenged the UK government's weakening carbon policy, also hopes the new generation will alter stale politics. (Radio NZ, 10 February 2024).

We refuse to believe we could learn anything from savages - "...the wise ones have an unvarying message to those white trespassers who have treated them so ill: the Earth is in peril; moderate your behaviour and help maintain it in the condition that we inherited, long before you came." (Winchester 2023).
Like C. H. Waddington "Tools for thought" 1977, I realise "if one does something to make an alteration in a complex system, the response of the system may not at all be what was expected or intended. This is a lesson which people continually have to learn over again". Hence I am resigned to the extinction of humanity (probably as a result of non-ecological scientists trying to 'fix it') and find no reason, nor even necessity, for hope, agreeing with Aristotle: "Hope is a waking dream". Although pessimistic I am not despondent; indeed, an early extinction would preserve more of what I find most beautiful on earth, and try to protect.

The American wit Woody Allen warns us, "...mankind faces the crossroads...one path leads to despair and utter hopelessness, the other to total extinction. I pray we have the wisdom to choose wisely." Perhaps I choose both: Frost's road not taken, and the road less travelled; rather than Gould's more hopeful version (1987) "I will cast my lot for a diversity of options - for our complex world may offer many paths to salvation, and the hounds of hell press continually upon us".

Mary Oliver, a fine environmental poet, writes: "When death comes...I don't want to end up simply having visited this world." Richard Dawkins, too, recommends we use our time well: "Within decades we must close our eyes again. Isn't it a noble, an enlightened way of spending our brief time in the sun, to work at understanding the universe and how we have come to wake up in it?"

Don Marquis, writing as Archie the typing cockroach (a poet reincarnated) pointed out our problem as long ago as 1927 in "what the ants are saying". After listing 12 civilisations that ended in deserts, Archie concludes:

"men talk of money and industry
of hard times and recoveries
of finance and economics
but the ants wait and the scorpions wait
for while men talk they are making deserts all the time
getting the world ready for the conquering ant
drought and erosion and desert
because men cannot learn."

Nearly 80 years later, in 2005, Jared Diamond expanded this topic with a 592 page book "Collapse: How societies choose to fail or succeed". And we had been advised long ago by Marcus Aurelius (Stoic philosopher, and Roman Emperor 161-180 AD) "Look back over the past, with its changing empires that rose and fell, and you can foresee the future too".

Few people realise, even now, that we have probably already failed:
"...the overall process of mass extinction cannot be undone... the fact that a pandemic [Covid-19] arises in a situation where 96% of planet's mammalian biomass belongs to humans and their livestock is not a coincidence." Christian Baron, 2022, "A planetary challenge on an unprecedented scale: the Sixth Mass Extinction is underway - and it represents a threat to human societies in more ways than one." Farsight, Futures for the Living World. Copenhagen Institute for Futures Studies, July No 02.

And Wilson, who spent the winter October 1888 - May 1899 at Davos in Switzerland being treated for tuberculosis, would be sad to know the city hosted an international conference concluding:

"Davos 2024 set the stage for global cooperation reshaping economic paradigms, navigating AI's evolution, and urgently addressing climate crises - endeavors crucial for safeguarding the Amazon and promoting a sustainable future." Woolly words, to delay actual action.

Hope

However, like Gould, I realise that some readers may find themselves "discouraged at the prospect of life as a detail in a vast universe" and worry about losing all hope.
It is up to you: "The universe doesn't owe you a sense of hope" (Richard Dawkins).

Visit the natural world. Orsolya Haarberg, 2017, introducing some of the finest photography I have ever seen in a book, "Laponia: Majestic Stillness", writes:

"A sense of stillness is much more than a mere absence of sound. There are sounds that feel like silence, soothing sounds, like those of nature. Listening to the whistling of the wind, the trickling of a brook, birdsong in a quiet forest, we unconsciously assimilate the rhythms of nature, that in turn, have a profound impact on our well-being."

And John Ruskin: "Nature is painting for us, day after day, pictures of infinite beauty" as Van Gogh knew, "If you truly love nature, you will find beauty everywhere".

Listen to the "Prayer of St Francis of Assizi, Let me be a channel of your peace," sung by Sinead O'Connor in 1997 - "Let me bring hope". (Probably written by Fr E. Bouquerel, 1912.)

Do not listen to Gustav Mahler's Das Lied von der Erde (The song of the Earth):
"...a persistent message that 'The earth will stay beautiful forever, but man cannot live for even a hundred years'."
"Mahler also hesitated to put the piece before the public because of its relentless negativity, unusual even for him. 'Won't people go home and shoot themselves?' he asked." (Wikipedia).

Read "The Book of Hope: A Survival Guide for Trying Times" by Jane Goodall and Douglas Abrams, 2022. This is by far the most inspiring book on the topic I have found, and the format of thought-provoking conversations, intriguing examples, mysticism, and toleration of other views is absolutely fascinating.

Percy Bysshe Shelley, in "To Maria Gisborne in England, from Italy", asks if they will meet again-

> "and she replies, Veiling in awe her second-sighted eyes; I know the past alone - but summon home My sister Hope, - she speaks of all to come."

Albert Einstein advises us to widen "our circle of compassion to embrace all living creatures and the whole of nature in its beauty... Look deep into nature, and then you will understand everything better". Each detail is equally important: value friendship, social and environmental diversity, be tolerant; seek, and find, beauty in everything - as Wilson did. Scott's last letter to Mrs Wilson, while they were dying, says "His eyes have a comfortable blue look of hope..."

Elusive butterflies:
by chance showing a transfer of
pattern from the old insect,
where they meet.

As for this book,

"Don't be concerned, it will not harm you
It's only me pursuing something I'm not sure of
Across my dreams, with nets of wonder
I chase the bright elusive butterfly of love."

Elusive Butterfly, Bob Lind (1965).

The Greeks "gave the same name, Psyche to the soul, or spirit of life, and to the butterfly" (W. S. Coleman, 1867, "British Butterflies") - Wilson probably had a copy of this book, which has superb colour illustrations of every species, but not the monarch which only recently became a rare vagrant there. However, although he should have seen it in South Africa, Australia, and New Zealand, as one of the commonest species, his diary (Wilson, E. 1972) mentions no butterflies at all. But in Bass Strait they struck thousands of moths: "The whole sea was covered with them and the ship was full of them, the air also..." From his description "like yellow under wings but all brown and grey" they were clearly bogong moths *(Agrotis infusa),* by coincidence "notable for its biannual long-distance migrations towards and from the Australian Alps, similar to the diurnal monarch butterfly." en.wikipedia.org. Both are now on the IUCN endangered species list.

Another Psyche, the lover of Cupid, is often drawn with a butterfly over her head.

Butterflies have been regarded as spiritual messengers or reincarnations in many First Nations, in Africa, North and South America, Asia including China and Japan, Australia, and Europe. And "The butterfly has symbolised ephemerality - and rebirth - for millennia" - Mathew Wilson, BBC, 16th September 2021.

Homero Aridjis, a Mexican politician and poet who helped preserve the monarch butterflies wintering forests from logging, said locals regard them as returning family spirits. His famous poem, "To a Monarch Butterfly" ends:

"...tell me what supernatural life is painted on your wings
so that after this life I may see you in my night."

(One ability of monarchs that remains "magic" is that migration north from Mexico to Canada takes three or four generations; the fourth or fifth generations know to fly south, on the longest butterfly migration known, and how to locate their winter trees in Mexico. Bogong moths likewise locate traditional caves they have never seen, and shelter at densities of 17,000 per square metre.)

In 2008 I attended a conference in Morelia, Mexico, and the optional field trip was to see the monarchs. By coincidence Gavin Van Horn, in his 2018 book "The Way of Coyote", also describes attending a conference in Morelia with a field trip to the monarchs, resulting in a chapter in his book, ending: "Monarchs are speaking to us, if only by their dwindling numbers, about the health of our commonly shared environment. Let us see if our respective journeys toward a life-giving future can be conjoined."

Jane Goodall, at her Gombe research area, found "We humans, with our passion for defining things, have named that spark in ourselves as our soul or spirit or psyche. But as I sat there, embraced by all the wonder of the forest, it seemed that that spark animated everything from the butterflies that fluttered past to the giant trees with their garlands of vines".

"Once Zhuang Zhou [a Chinese Daoist philosopher, 400 B.C. I dreampt he was a butterfly ... fluttering around, happy with himself and doing what he pleased... he woke... solid and unmistakably Zhuang Zhou. But he didn't know if he was Zhuang Zhou who had dreampt he was a butterfly, or a butterfly dreaming he was Zhuang Zhou. Between Zhuang Zhou and a butterfly there must be some distinction! This is called the Transformation of Things." (Burton Watson, 2003).

Final coincidences

Dr John W. Etheridge (1804-1866), a famous religious scholar and family relation - whose surname I was given as a middle name - translated the New Testament from Syriac.

Wilson also translated this, spending a year, before the other residents at the Caius mission were awake, on "a paraphrase of almost the whole of the New Testament, committed to paper to clarify and record the meaning for himself" (Seaver 1948).

Etheridge's copperplate handwriting, on unruled paper, is an art worth sharing: one of 11 translations of the same verse from languages he knew, written by quill pen in his daughter Eliza's 12cm wide autograph album:

Psalm XVII : 15.

As for me, I will behold thy Face in Righteousness : I shall be satisfied when I awake with thy Likeness

Hebrew

אֲנִי בְּצֶדֶק אֶחֱזֶה פָנֶיךָ אֶשְׂבְּעָה
בְהָקִיץ תְּמוּנָתֶךָ׃

Ani, betzedek echezeh panecha; esbeyah behakitz temunatecha.

I in righteousness will behold thy Face; I shall be satisfied in awaking (with) thine image.

And Etheridge, in a poem he wrote titled "Know'st Thou the Land?" offers this advice:

"Fair Psyche spreads her wings, plum'd for the flight,
Which bears her, deathless, to the climes of light;
Know'st thou it well? 'tis freedom's glorious shrine,
And thither must thou rise to find it thine."

(Published in Thornley Smith, 1871, "Memoirs of the Rev. John Wesley Etheridge..." Hodder and Stoughton, London.)

Envoi

Francis Spufford (1996) in "I May be Some Time. Ice and the English Imagination" (faber and faber, London) gives a most nuanced account of the great polar explorers' lives, characters, and deaths. "Cherry-Garrard writes in this diary that Wilson 'had died very quietly', and Bowers 'also quietly'. He does not care to guess how Scott died...'It is all too horrible'" - but Glyn Maxwell, reviewing Spufford's book for Independent on Sunday, does: "Scott hearing their breath turn ragged and end, while he wrote on in what must have been about the most abominable loneliness known to a human being."

Spufford concludes,"They collapse the tent gently. They build a cairn over it that stands black against 'sheets of iridescent clouds'. And they turn away; so shall we".

Leaving "sheets of iridescent clouds...a glory of burnished gold" Apsley Cherry-Garrard.

Acknowledgements

I am grateful to David Wilson and Nicholas Reardon for their encouragement and advice on producing this book. Sincere thanks to Dr Mike Rudge, who critically reviewed an early draft, re-ordered sections to read more clearly, and clarified my confused ideas; to Professor Kim King for her flattering editorial, and for reminding me of Hamlet. Many friends watched slide-shows and gave helpful comments. Sharon Murdoch very kindly allowed me to use my favourite cartoon character, Munro the crossword cat. Much of the text is direct quotation, and I am glad to acknowledge the brilliant writers who have expressed these ideas far better than I ever could. Not forgetting the coincidence of treecreepers which led me to Uncle Bill, and a new appreciation of life, death, and psyche. As Frank Debenham wrote to Scott's widow on 6 November 1913, "Life has become at once simpler and more real".

References

Abrahams, M. 2023. Spacey superpowers. New Scientist, 20 April 202, No. 3435: 56.

Attenborough, D. 2010. The Living Planet: A Portrait of the Earth. The augmented and enlarged edition. Published by Reader's Digest, William Collins, and the BBC, London. 368pp.

Bach, R. 1977. Illusions: The Adventures of a Reluctant Messiah. Dell, New York. 192pp.

Baker, R.R. 1980. Goal Orientation by Blindfolded Humans After Long-Distance Displacement: Possible Involvement of a Magnetic Sense Science 210: 555-557.

Ball, P. 2017. Demographics. Pp. 7-19, In Jim Al-Khalili, Ed.,"What's Next?" Profile Books, London, 250pp.

Baron, C.K. 2022. A planetary challenge on an unprecedented scale: the Sixth Mass Extinction is underway - and it represents a threat to human societies in more ways than one. Farsight, Futures for the Living World. Copenhagen Institute for Futures Studies, July No 02: 24-31.

Barrett-Hamilton, G.E.H. 1912. A History of British Mammals: with twenty-seven full-page plates in colour, fifty-four in black and white, and upwards of two hundred and fifty smaller illustrations drawn by Edward A. Wilson. 3 Vols. Gurney and Jackson, London.

Bechly, G. 2023. Fossil Friday: A Dinosaur Feather and an Overhyped New Study on the Origin of Feathers. Evolution News & Science Today, 26 May 2023.

Beitman, B. 2022. Meaningful Coincidences: How and Why Synchronicity and Serendipity Happen. Park Street Press, Rochester, Vermont. 208pp.

Butfield, C. and J. Hughes. 2021. Earthshot: How to Save our Planet. John Murray, London. 338 pp.

Cherry-Garrard, A. 1965. The Worst Journey in the World: Antarctic 1910-13. New Edition. Penguin Books in association with Chatto & Windus, Auckland. 652pp.
Coleman, W.S. 1867. British Butterflies. George Routledge & Sons, London. 179pp.
Dawson, D.G. and P.C. Bull. 1975. Counting birds in New Zealand forests. Notornis 22: 101-109.
Descartes, R. 1637. Discourse on Method and the Meditations. Trans F.E. Sutcliffe, 1968. Penguin Books, London. 188pp.
Diamond, J. 2005. Collapse: How societies choose to fail or succeed. Penguin Books, London. 592pp.
Dubos, R. 1970. So human an animal. ABACUS, Sphere Books, London. 192pp.
Dunn, R. 2022. A natural history of the future: What the laws of biology tell us about the destiny of the human species. Basic Books, London. 306pp.
Dunne, J.W. 1927. An Experiment with Time. 1934 Ed. Faber & Faber Ltd., London. 208pp.
Flannery, T. 2015. Atmosphere of Hope: Searching for solutions to the climate crisis. Text Publishing, Melbourne. 245pp.
Fiennes, R. 2003. Captain Scott. Hodder & Stoughton, London. 508pp.
Fleming, P. 1933. Brazilian Adventure. Penguin Books edition, 1984, Auckland, NZ. 373pp.
Flux, J.E.C. 1958. Red grouse in autumn: Age and sex characteristics studied.The Scottish Landowner No. 91: 43-44.
Flux, J.E.C. 1965. Incidence of ovarian tumors in hares in New Zealand. Journal of Wildlife Management 29: 622-624.
Flux, J. 2016. Illustrate Ecology: Fertiliser to fertiliser. NZ Ecological Society Newsletter 156:1.
Flux, J.E.C. and M. Flux. 2015. The fertility clinic: a bird's-eye view of our future. New Zealand Journal of Zoology, DOI: 10.1080/03014223.2015.1099549
Flux, J.E.C. and M. Flux. 2018. Deevey's Hare and Haruspex revisited: Why domestication dooms civilisation? European Journal of Ecology, 4(2): 100-110.
Flux, J.E.C. and M. Flux. 2019. Polymorth stability, and changed flight period, of *Declana floccosa* Walker, 1858 (Lepidotera: Geometridae) in New Zealand, 1974-2016. New Zealand Entomologist, 42(2): 100-109.
Foell, E.W. and R.A. Nenneman. 1986. How Peace Came to the World. The MIT Press, London. 257pp.
Fox, D. 2022. The coming collapse. Scientific American, 327, 5: 24-33.
Goodall, J., D. Abrams, G. Hudson. 2022. The Book of Hope: A Survival Guide for Trying Times. (Viking, 2021) Penguin Random House UK. 249pp.

Gould, S.J. 1985. The Flamingo's Smile: Reflections in Natural History. Norton and Co., New York. 476pp.

Gould, S.J. 1987. Freudian Slip. Natural History, 96: 14-21.

Gould, S.J. 1990. An Urchin in the Storm: Essays about Books and Ideas. Penguin, London. 255pp.

Gould, S.J. 2002. Rocks of Ages: Science and religion in the fullness of life. Vintage, Random House, London. 241pp.

Grandin, T. and C. Johnson. 2006. Animals in Translation: Using the Mysteries of Autism to Decode Animal Behaviour. Bloomsbury Publishing, London. 356pp.

Grayling, A.C. 2023. Philosophy and Life: Exploring the great questions of how to live. Viking, Penguin Random House, UK. 426pp.

Guiley, R.E. 1991. Harper's Encyclopedia of Mystical & Paranormal Experience. HarperCollins Publishers, New York. 666pp.

Haarberg, E., O. Haarberg, J. E. Utsi. 2017. Laponia: Majestic Stillness. Jokkmokk Sami Kompania. Jokkmokk. 168pp.

Herman, A. 2001. The Scottish Enlightenment: The Scots' Invention of the Modern World. Harper Perennial, London. 454pp.

Huxley, E. 1977. Scott of the Antarctic. Weidenfeld and Nicholson, London. 303pp.

King, C.M. 2015. Editorial. Introduction to "The fertility clinic: a bird's-eye view of our future". New Zealand journal of Zoology, DOI: 10.1080/03014223.2015.1099550

Kirschvink, J.L., A. Kobayashi-Kirschvink, J.C. Diaz-Ricci, and S.J. Kirschvink. 1992. Magnetite in human tissue: a mechanism for the biological effects of weak ELF magnetic fields. Bioelectromagnetics Supplement 1: 101-113.

Leslie, A.S. and A.E. Shipley. 1912.The Grouse in Health and Disease. (Popular Edition). Smith, Elder and Co., London. 472pp.

Marquis, D. 1931. Archy and Mehitabel. 1957 edition, Faber & Faber Ltd. London, 166pp.

Macan, T.T. 1974. Freshwater Ecology. Second edition, Longman, London. 343pp.

Mallove, E. 1985. Einstein's Intoxication With the God of the Cosmos. Washington Post, 22 December 1985.

Pasternak, B. 1962. In the Interlude: Poems 1945-1960. Trans. H. Kamen. Oxford University Press, Oxford. 143pp.

Pound, R. 1966. Scott of the Antarctic. 1968 Edition, World Books, London. 304pp.

Ritchie, H. 2024. Not the End of the World: How We Can Be the First Generation to Build a Sustainable Planet. Chatto & Windus, London. 340pp.

Ripple, W.J. *et al.* 2017. World Scientists' Warning to Humanity: A Second Notice. Bioscience 67: 1026-1028.

Rousseau, J.-J. 1782. Meditations of a solitary walker. Peter France trans., 1979. Penguin Books, London. 54pp.

Rowthorn, A. 1989. Caring for Creation:Toward an ethic of responsibility. Morehouse Publishing, Connecticut. 163pp.

Seaver, G. 1936. Edward Wilson of the Antarctic: Naturalist and friend. John Murray, London. 301pp.

Seaver, G. 1937. Edward Wilson Nature-Lover. John Murray, London. 219pp.

Seaver, G. 1948. The Faith of Edward Wilson. John Murray, London. 48pp.

Sender, R., S. Fuchs, and R. Milo. 2016. Revised Estimates for the Number of Human and Bacteria Cells in the body. PLoS Biology, 14(8): e1002533.

Smith, T. 1871. Memoirs of the Rev. John Wesley Etheridge, M.A., PH.D. including extracts from his writings, correspondence, and poetry. Hodder and Stoughton, London. 323pp.

Spedding, J., R.L. Ellis, and D.D. Heath. 1858. The Works of Francis Bacon. Vol.V. Translations of the Philosophical Works, Vol. 11. Longman and Co., London. 682pp.

Spufford, F. 1996. I May be Some Time: Ice and the English Imagination. faber and faber, London. 372pp.

Sukel, K. 2023. Colourful thinking. New Scientist Vol. 258, Issue 3442: 40-43.

Stevens, C., G. O'Connor, and N. Robinson. 2019. The connections between art and science in Antarctica: Activating Science*Art. Polar Record, 55(4): 1-8.

Suzuki, D. with A. McConnell. 1997. The Sacred Balance: Rediscovering our place in nature. Allan and Unwin, Australia. 262pp.

Swinton, W.E. 1977. Physicians as explorers. Edward Wilson: Scott's final Antarctic companion. CMA Journal, 117: 959-974.

Thomas, L. 1979. The Medusa and the Snail: More notes of a biology watcher. (The Viking Press) Penguin Books, 1995, London. 175pp.

Thomas, L. 1984. Late Night Thoughts. Oxford University Press, UK. 175 pp.

Van Horn, G. 2018. The Way of Coyote: Shared Journeys in the Urban Wilds. The University of Chicago Press, Chicago. 234pp.

Waddington, C.H. 1963. The Nature of Life. Allen and Unwin, London. 128pp.

Waddington, C.H. 1977. Tools for Thought. Paladin, Frogmore, St Albans, Herts. 250pp.

Warne, K. 2017. Viewpoint. Edward Wilson's Bird. New Zealand Geographic. http://www.nzgeo.com/stories/edward-wilson-bird/

Watson, B. trans. 2003. Zhuangi: Basic Writings. Columbia University Press, New York. 168pp.
Watson, R. 2011-12. The art and science of ice: Edward Adrian Wilson's ice crystal drawings. PCAS 14. 19pp.
Wilson, D.M. and D.B. Elder. 2000. Cheltenham in Antarctica: The life of Edward Wilson. Reardon Publishing, Cheltenham. 144pp.
Wilson, D.M. and C.J. Wilson 2004. Edward Wilson's Nature Notebooks. Reardon Publishing, Cheltenham. 168pp.
Wilson, D.M. and C.J. Wilson 2011. Edward Wilson's Antarctic Notebooks. Reardon Publishing, Cheltenham. 184pp.
Wilson, E. 1972. Diary of the Terra Nova Expedition to the Antarctic 1910-1912. Ed. H.G.R. King. Humanities Press, New York. 279pp.
Winchester, S. 2023. Knowing what we Know: The transmission of knowledge from ancient wisdom to modern magic. William Collins, London. 415pp.
Witherow, D. 1981. The real dilemma: when compromise is cop-out. NZ National Business Review, 19 October, 1981.
Wynne-Edwards. V.C. 1962. Animal Dispersion in relation to Social Behaviour. Oliver and Boyd, Edinburgh and London. 653pp.

List of Illustrations and Copyright Acknowledgement

Songs and Poems are quoted in conformation with the legal term "Fair dealing". This advises the shortest necessary amount, and no loss of revenue to the owner. For Archy by Don Marquis, © by the owner, provided at no charge for educational purposes.

Wilson's images are widely dispersed, and held under various forms of copyright. The full name of the holder is given on first mention, and the form of recognition as requested.

With thanks to:

1 © AHT (Antarctic Heritage Trust). Dr Edward Wilson's watercolour entitled "Treecreeper" discovered at Carston Borchgrevink's hut at Cape Adare. *Watercolour-PhaseONE_Stain Removed.*
2 © NHM (Natural History Museum, London) T36211. Colour change in Scottish mountain hares. A History of British Mammals, Barrett-Hamilton (1912). Public Domain.
3 © The Grouse in Health and in Disease: being the popular edition of the report of the Committee of Inquiry on Grouse Disease, by A. S. Lesley and A.E. Shipley (1912). (Copyright-evidence-operator, Liz Ridolfo, "no visible notice of copyright").
4 © The Grouse in Health and in Disease, Lesley and Shipley (1912). As above.
5 © private. Swift head.
6 © private. Magpie.
7 © private. Sparrow hawk.
8 © private. Swallow.
9 © private. Grey wagtail.
10 © private. Sea trout.
11 © private. Char.
12 © NHM. Fish, Agulhas Bank, SA. 15 Oct. 1901. Public Domain.
13 © CAGM. (Cheltenham Art Gallery and Museum). 1995.550.171. Writing with pen.
14 © CAGM. 1995.550.171. Fists.
15 © private. Moths.
16 © private. Grass snake.
17 © private. Mosquitos.
18 © SPRI (Scott Polar Research Institute) P69/10/1-1880 and 1881 Butterflies.
19 © private. Lizard.

20 © NHM 1454 387. Vole. Public Domain.
21 © NHM T36137. Bat. Public Domain.
22 © private. Cancer.
23 © CAGM 1995.550.171. Dog skull.
24 © NHM z88fw. Dolphin. Public Domain.
25 © NHM 058526. Dolphin. Public Domain.
26 © NHM T36217. Seals. Public Domain.
27 © NHM T36109. Seal sketches. Public Domain.
28 © NHM T36213. Orcas. Public Domain.
29 © NHM T36154. Hares. Public Domain.
30 © NHM T30288. Rabbits. Public Domain.
31 © NHM T56238. Rabbits. Public Domain.
32 © Alami stock photo. Hares.
33 © NHM T36119. Stoat. Public Domain.
34 © NHM T36138. Roe buck. Public Domain.
35 © NHM T36139. Roe deer. Public Domain.
36 © NHM T36165. Fallow deer. Public Domain.
37 © private. Reindeer.
38 © private. Beetle.
39 © private. Flies.
40 © private. Ladybird.
41 © private. Wasp nest.
42 © NHM T36207. Shrew. Public Domain.
43 © private. Newts.
44 © The Grouse in Health and in Disease, Lesley and Shipley (1912).
45 © The Grouse in Health and in Disease, Lesley and Shipley (1912).
46 © SPRI Y: 67/4/3. sea leopard.
47 © NHM, Tern, from Seaver,1936. Public Domain.
48 © CAGM, 1995.550.171. Turkey.
49 © private. Brent goose.
50 © private. Lapwing.
51 © private. Turnstone.
52 © private. Redshank.
53 © private. Golden plover.

54 © SPRI. N:1838. Rooks.
55 © SPRI. N:1843. Chough.
56 © NHM. T36180. Wild cat. Public Domain.
57 © SPRI. N:1847. Sparrows.
58 © CPL (Cheltenham Public Library). 1881. Titlark.
59 © private. Woodpecker.
60 © private. Starlings.
61 © SPRI. N: 192. Whitethroat.
62 © private. Hoopoe.
63 © SPRI. N:1825. Oystercatcher.
64 © SPRI. N:1468. Ptarmigan.
65 © Mary Evans Picture Library / Natural History Museum, London. Polecat.
66 © private. Knot.
67 © SPRI. N:1807. Ringed plover.
68 © CC (Cheltenham College). 002. Flying birds.
69 © private. Heron.
70 © private. Hazel grouse.
71 © SPRI. N:1870. Chaffinch.
72 © SPRI. N:1861. Redpoll.
73 © private. Siskin.
74 © private. Goldcrest.
75 © private. Long-tailed tit.
76 © SPRI. N:1884. Crossbill.
77 © private. Grey partridge.
78 © private. Primrose.
79 © private. Mountain avens.
80 © private. Bramble/Blackberry flower.
81 © private. Bramble/Blackberry fruit.
82 © private. Gentian.
83 © private. Moss campion.
84 © private. Violets.
85 © private. Anemone.
86 © private. Large wintergreen.
87 © private. Larch cones.

88 © private. Red fungus.
89 © private. Toadstool cut.
90 © private. Tree with owl.
91 © private. Honeysuckle.
92 © CAGM. 1995.550.170. Spires.
93 © CAGM. 1995.550.164. Cathedral interior.
94 © CAGM. September 1896. Walls and chimneys.
95 © CC. 152a. Dingle beach.
96 © AHAG (Abbot Hall Art Gallery and Museum, Kendal). 02850/87.
"Sunset, McMurdo Sound, April 13.11 5pm", watercolour by Edward Adrian Wilson, 1911.
Given to Abbot Hall in 1987, reproduced by courtesy of Abbot Hall, Lakeland Arts Trust.
97 © SPRI. N: 433. Clouds.
98 © SPRI. N: 452. Clouds.
99 © SPRI. N: 1281. Calm sea.
100 © SPRI. MS900/2 - Fairy wren.
101 © SPRI. MS900/2 - Tawny frogmouth (sketch).
102 © SPRI. MS: 234/4.Volcano, NZ
103 © SPRI. Y: 2006/15/41c Maori Whare.
104 © SPRI. N: 1416. Sea lions.
105 © private. Sea lion male.
106 © SPRI. N: 1401. Ice cave.
107 © SPRI. N: 1819. Dipper.
108 © SPRI. N: 1880. Coal tit.
109 © private. Black-headed gull.
110 © private. Great skua..
111 © private. Cat expressions
112 © SPRI. N: 471. Sun halo.
113 © SPRI. N: 1634. Albatross.
114 © SPRI. N: 1629. Composite birds.
115 © SPRI. N: 421. Albatross.
116 © SPRI. MS234/4. Adelie penguin (sketch).
117 © SPRI. N: 1937. White-eye.
118 © SPRI. N: 1722. Tern.
119 © CAGM. 1995.550.171. Heads.

120 © private. Pukeko.
121 © private. Lapwing.
122 © SPRI. N: 65/3/1896. Mice.
123 © SPRI. N: 475. Volcano.
124 © private. Misty hills.
125 © SPRI. N: 1388. Spiral cloud.
126 © SPRI. N: 1935. Falcon.
127 © SPRI. N: 1671. Prion.
128 © SPRI. N: 1697,1698,1699. Cape pigeon.
129 © private. Frigate.
130 © DHT (Dundee Heritage Trust). South Polar Times, June,1903. Dogs.
131 © SPRI. N: 1816. Goosander.
132 © SPRI. N: 1801/58. Ice crystals.
133 © Cheffins. Tubularia.
134 © DHT. South Polar Times, June, 1902. Scale worm.
135 © NMBL (National Marine Biological Library) 18 June 1903. Chiton.
136 © DHT. South Polar Times, June, 1902. Isopod.
137 © DHT. South Polar Times, April, 1902. Met equipment.
138 © SPRI. N: 1720. White tern.
139 © SPRI. N: 1847. Hooded crow.
140 © SPRI. N: 1731. Frigate.
141 © SPRI. MS234/4 - Adelie penguin (sketch).
142 © private. Scoter.
143 © private. Duck sketch.
144 © SPRI. N:1865. Wigeon.
145 © private. Black guillemot.
146 © SPRI. N:1835,1836. Gannet.
147 © NHM T36212. Notebook. Public Domain.
148 © private. Robin, end page.
149 © CL (Cheltenham Library) Diving.
150 © CAGM 1995.550.171. Portrait.

About the Author

John Flux

Born 1934 in Maymyo, Burma; moved to Channel Islands, UK, 1939 -1940; Scotland 1940-1960. At school (Mackie Academy, Stonehaven) no biology was taught, but in the final year 2 hours every Friday was spent on Hamlet. After a BSc and PhD at Aberdeen university, John joined Ecology Division DSIR in 1960 to study hares at night. To occupy the day, a starling study began in 1970. Apart from papers on these, as an old-style naturalist he has published on trout, grouse, swallows, thrushes, blackbirds, wallabies, butterflies, cats, pigeons, possums, rats; fleas, ticks, and seed burrs carried; leaf shapes and colour, tree-trunk diameter, by-the-wind sailors, bird-nest/plant mutualism, dandelions, lichens, wool-carder bees, harriers on road-kill, colour-blindness, moth evolution, bird spacing, and "Biogeographic theory and the number and habitat of moas".

Photograph by Merryl Park

BES - #0037 - 130824 - C161 - 210/235/20 [22] - CB - 9781901037272 - R - HT - Matt Lamination